Case Studies in Post-Construction Liability and Insurance

T0330964

Case Studies in Post-Construction Liability and Insurance

Edited by
Anthony Lavers
Centre for Real Estate Management
Oxford Brookes University, UK

LONDON AND NEW YORK

First published 1999 by Taylor & Francis

2 Park Square, Milton Park, Abingdon, Oxfordshire OX14 4RN
52 Vanderbilt Avenue, New York, NY 10017

Routledge is an imprint of the Taylor & Francis Group, an informa business

First issued in paperback 2019

British Library Cataloguing in Publication Data
A catalogue record for this book is available from the British Library

Library of Congress Cataloging in Publication Data
A catalog record for this book has been requested

ISBN 978-0-419-24570-4 (hbk)
ISBN 978-0-367-39964-1 (pbk)

Publisher's Note
The publisher has gone to great lengths to ensure the quality of this reprint but points out that some imperfections in the original may be apparent.

Contents

Preface

The spirit in which this collection of Case Studies in Post-Construction Liability and Insurance was made is that on which W87 has always operated, namely:

None of us has a monopoly of wisdom. Since no country knows all the answers as to how to resolve these costly, time-consuming and often traumatic disputes, even within its own borders, it is taken as self-evident that all can benefit from knowing more of the experiences of others in addressing essentially similar issues. Those experiences will include failures and successes; it is possible to learn from both.

The intention of W87 and TG15, now W103 in undertaking this work was to make available to a wider audience, specifically to the research and professional communities of the construction industry world-wide, the knowledge of its members and the colleagues and contacts who have assisted them.

My task as Editor has been a humble one, in the sense that the credit for the Case Studies belongs exclusively to the authors who wrote them. Insofar as my introductory and commentary chapters help to illuminate themes, I am glad to have had the opportunity of learning from those authors. If I have been guilty of misunderstanding or wrong emphasis in construing their messages, I apologise – the best guide for readers will be to go beyond my points and into the Case Studies themselves.

Acknowledgements

It is literally impossible to acknowledge by name every person who has contributed to a work of this kind. Every member of the two Working Commissions 87 and 103 has contributed in some way, directly or indirectly. It is obvious that the authors have had assistance from colleagues with specialist knowledge to offer; they have indicated this. I am grateful to all.

The first specific thanks must go to the authors. They have worked very hard in research and in writing, then in revising and dealing with further queries. They had to produce their contributions often against extremely demanding professional commitments. Many (although not all) still managed to meet deadlines and (usually) respond to communications. They were also (mainly) understanding during the frustrating delays which have occurred, often at my end, during the process of collection, editing and submission for publication. I hope that they will consider the finished product an adequate recompense for their skill, effort and patience.

I should also like to thank those people, generally members of the Commissions, who acted as 'focal points' for their countries. These people, sometimes authors, sometimes not, helped to co-ordinate the responses.

Several people helped me through the early stages of collection and editing. I acknowledge the contributions of Sarah Marsh and Mendel Bickovsky. Catherine Hoar, Secretary to the Centre for Real Estate Management at Oxford Brookes University deserves a special mention; she helped a great deal. The University helped me in a number of other ways with facilities and funding.

I should like to thank my opposite numbers in W103, Peter Fenn and Edward Davies for their co-operation and that of their members; the dispute resolution perspective makes the case studies more complete.

In getting over fifty contributions ready and together, Mrs. Alexandra Rivers has been a tower of strength at the word processor, whose ability to keep working with care and cheerfulness was a great support.

Tim Robinson of E&FN Spon has been positive and reassuring on behalf of the publishers.

Dr Alicia Montarzino was entirely responsible for securing the first contributions to our work from the South American sub-continent. She used her knowledge, network of contacts and language skills to make possible involvement of experts from Argentina, Chile and Uruguay, which could otherwise not have happened.

I conclude these lengthy but deserved acknowledgements by reference to a remarkable person, known to many of the contributors to this book. Jens Knocke created W87 and previously showed the way to publish its work, as well as his own. Where this book falls short of his exacting standards, it is entirely my responsibility. That it has appeared at all is due to his enormous help in the sub-editing and manuscript preparation processes. His gifts of intellect, linguistic brilliance and facility with information technology, coupled with his unique personal qualities, entitle him to a normally misused title. Jens is indeed a great man.

About the Editor

Anthony Lavers is Professor of Law in the Centre for Real Estate Management at Oxford Brookes University. He is a Barrister (Lincoln's Inn) and is Consultant to the Construction Group at London solicitors Barlow Lyde and Gilbert. He has been a member of CIB W87 since 1986 and became its Co-ordinator in 1994. Anthony Lavers was elected a Member of Council of the Society of Construction Law in 1995 and re-elected in 1997.

Contributors

Argentina	Professor Ing. **Isaac E. Edelstein** Faculty of Architecture National University of Córdoba Córdoba
Australia	**Deepak Bajaj** Department of Construction Management University of Technology, Sydney **Cliff Roberts** Associate Professor, Faculty of Design, Architecture and Building University of Technology, Sydney **Stephen Smith** Barrister, Melbourne
Canada, Quebec	**Laurent Arsenault** Engineer, Montral Maître **Daniel Alain Dagenais** Lavery de Billy, Avocats, Montréal
Chile	**Christian Canales Palacios** Attorney, Santiago **Juan Claudio Molina** Attorney, Santiago **Gastón Escudero Poblete** Attorney, Santiago **Juan Carlos León Flores** Civil Engineer, Santiago **Alberto Ureta Alamos** Civil Engineer, Santiago

China

Deepak Bajaj
Department of Construction Management
University of Technology, Sydney, Australia
Zhong Li
Shanghai Zhu Yi Construction Development
Ming Rong Shi
Jones Lang Wootton
Shanghai
John Twyford
Lecturer
Faculty of Design, Architecture and Building
University of Technology, Sydney, Australia
Rui Zhang
Faculty of Design, Architecture and Building
University of Technology, Sydney, Australia

Denmark

Advokat **Jens Jordahn**
Plesner & Grønborg
Copenhagen

England and Wales

Deborah Brown
Trainee Solicitor
Construction Group
Barlow, Lyde and Gilbert, London
Roy Pembroke
Solicitor
National House Building Council, Amersham
Patrick Perry
Solicitor
Construction Group
Barlow, Lyde and Gilbert, London
Caroline Pope
Solicitor, Partner
Construction Group
Barlow, Lyde and Gilbert, London
Roger Wakefield
Solicitor, Partner
Construction Department
Nabarro Nathanson, London

France	**François Ausseur** Direction Générale SMAbtp, Paris **Jean Desmadryl** Inénieur en Chef Paris
Germany	**Knut Richter** Rechtsanwalt Hannover **Regina Tuscher** Vereinigte Haftpflichversicherung Hannover
Hong Kong	**Edwin Hon-wan Chan** Associate Professor Department of Building and Real Estate Hong Kong Polytechnic University Dr **Mohan Kumaraswamy** Associate Professor Department of Civil Engineering University of Hong Kong **Robert Morgan** Barrister Hellings Morgan Associates, Hong Kong **Yoshiyuki Ueda** Architect General Affairs Department International Operations Division Tobishima Corporation, Japan **Kumaru Yogeswaran** Engineer Scott Wilson (HK) Ltd., Hong Kong
Japan	**Fumihiko Omori** Architect and Attorney Omori Law Office, Tokyo Professor **Kohei Matsumoto** Faculty of Real Estate Science Meikai University, Tokyo

Shuji Motoki
Manager, Planning and Research Dept.
Registration Organisation for Warranted Houses
Tokyo
Ikuo Watanabe
Liability Insurance Division
Fire and Casualty Underwriting Dept.
Yasuda Fire and Marine Insurance Co., Tokyo
Michiko Yamauchi
Japan Building Equipment and Elevator
Centre Foundation, Tokyo

Kuwait

Tawfiq Al Bahar
General Manager
Warba Insurance Co. S.A.K
Jasem M. Jumaian
Engineer
Grand Project Management Unit
Real Estate Department Investment Office, Safat

Norway

Advokat Dr. juris **Jan Einar Barbo**
Bugge, Arentz–Hansen & Rasmussen, Oslo
Hans-Jakob Urbye
Norwegian Council for Building Standardisation
Oslo

Singapore

Raymond Chan
Solicitor, Partner
Chan Tan and Partners
George Tan
Solicitor, Partner
Chan Tan and Partners

Spain

Dr **Manual Olaya Adán**
Instituto de Ciencias de la Contrucción Eduardo
Torroja, Madrid

Sweden

Kåre Eriksson
AB Bostadsgaranti, Stockholm
Advokat **Anders G Kleberg**
Advokatfirman Anders Kleberg, London
England
Lennart Lundahl
Trygg Hansa, Stockholm

Switzerland

Professor Dr **Viktor Aepli**
Attorney at Law
Fribourg Law School
Dr.iur. **Roland Hürlimann**
Attorney at Law, Partner
Schumacher Baur Hürlimann
Zürich

United States

Wilson Barnes
Construction Department
Southern Polytechnic State University
Marietta, Georgia
Ron Craig
Department of Civil and Building Engineering
Loughborough University
Loughborough, United Kingdom

Uruguay

Carlos Altoberro
Engineer
Banco Hipotecario del Uruguay
Montevideo
Aldo Lamorte Russomano
Architect
Sociedad de Certificación y Control
Montevideo
Dr **Dora Szafir**
Judge, Professor of Law
Universidad de la República
Montevideo

PART ONE

PART ONE

CIB Commission W87 on Post-Construction Liability and Insurance and its Work

CIB

The International Council for Research and Innovation in Building and Construction (CIB) is a world-wide organisation which promotes and oversees collection and exchange of information and facilitates research and its dissemination. It is essentially a network of over 5000 construction experts from a total of 525 members, including 350 institutions, in more than 70 countries. The members operate in over 50 Working Commissions and Task Groups in areas of ongoing concern to the construction industries of the world and their research communities. CIB's General Secretariat is based in Rotterdam, where a full-time staff deals with issues of membership, establishment of new Working Commissions and Task Groups and the administration of existing ones and, increasingly publication of their work. A Working Commission is formed to co-ordinate and sometimes undertake research in a specific area of construction-related activity. A Working Commission will be made up of construction professionals, academics and researchers from government and private sector institutions. A Task Group is similar in constitution, but will have a single focus on a particular objective, rather than the somewhat more comprehensive approach to a subject of a Working Commission. The life of a Task Group might be shorter if its objective is accomplished, although some develop into Working Commissions.

Each Working Commission or Task Group is headed by a Co-ordinator and meets regularly; the most active annually, at Plenary Sessions or Symposia hosted by members' institutions. Commission members present papers and update each other on developments in their own countries, as well as internationally. This often leads to publication of the results, either through CIB itself or through commercial publishers. This publication is part of a rapidly-developing joint venture between CIB and E&FN Spon, the specialist construction imprint of international publishers Routledge.

Case Studies in Post-Construction Liability and Insurance, edited by Anthony Lavers. Published in 1999 by E & FN Spon, 11 New Fetter Lane, London EC4P 4EE, UK. ISBN: 0 419 24570 7

CIB W87

CIB Working Commission W87 on Post-Construction Liability and Insurance (W87) was set up in 1985 under its Founder Co-ordinator Jens Knocke, who had proposed the idea to CIB in 1983. An exploratory meeting was held in Brussels and the scope of the Commission was established as the study of the provisions relating to liability for defects in buildings after the construction stage and, where appropriate, the insurance available to cover the risk of such liability.

The Commission comprises approximately 80 members from the following 25 countries: Argentina, Australia, Belgium, Canada, Chile, China, Denmark, Finland, France, Germany, Italy, Japan, the Netherlands, New Zealand, Norway, Portugal, Rumania, Russia, Singapore, South Africa, Spain, Switzerland, United Kingdom, United States and Uruguay.

The members come from a range of backgrounds, both in the sense of their disciplines and their employment: insurers, lawyers, construction professionals (especially architects), as well as personnel from building institutes, government departments and professional bodies/trade associations.

The Commission is run by a Steering Committee of 14 senior members from nine countries. Since its foundation, the Commission has met in Montreal (1985), London (1986), Copenhagen (1987), Paris (1988), Edinburgh (1989), Lisbon (1990), Las Palmas (1991), Copenhagen and London (1992), Stockholm (1993), Washington D.C. (1994), Tokyo (1995), Oslo (1996), Paris (1997) and Uppsala (1998), where an invitation was received to hold the 1999 meeting in Sydney. At the Plenary Sessions, one of the most popular agenda items is 'What's new in members' countries', during which members from each country present give short updates on developments within their own systems. These could be changes in the law, trends in practice or the availability of new insurance products. Overviews are also given of any relevant activity at regional level: in the European Union (EU), NAFTA, ASEAN or MERCOSUR. A major presentation is always given by the host country and by any countries which have not previously had members. Papers may also be presented on comparative aspects of the subject, such as the effect of insurance in inhibiting innovation in design or the role of maintenance requirements in ascertaining liability in members' countries. These are sometimes presented by invited experts, as well as by members.

THE GREEN BOOK

In 1993, the Commission published, through E&FN Spon, 'Post Construction Liability and Insurance' (The Green Book) edited by the then Co-ordinator, Jens Knocke (ISBN 0-419-15350-0). This contained monographs which were National Overviews of the post-construction liability and insurance systems of the following 16 countries : Australia, Belgium, Denmark, England and Wales and Northern Ireland, France, Italy, Japan, the Netherlands, Norway, Portugal, Quebec (Canada), Scotland, Singapore, Spain, Sweden and the United States of America. In addition, Jens Knocke, Daniel Alain Dagenais and Gérard Blachère respectively, provided chapters on key concepts in the subject, law and contract relating to post-construction liability and the role of national governments in risk reduction.

The National Overviews were obtained through carefully chosen methodological preparation and execution. Jens Knocke, as Co-ordinator, circulated to each country represented a detailed questionnaire in a standard format requesting the respondents to address approximately 50 questions. Some questions asked for information, while others invited description and discussion. W87 members from 16 countries responded and the result was successful in two important respects. First and most obviously, a detailed account became available of the post-construction liability and insurance arrangements of some of the major jurisdictions of the world. This was a significant contribution to international study and a useful reference tool for practitioners engaged in cross-border work. Second, the book established a taxonomy for the subject which enables many of the confusions and downright errors resulting from inconsistent use of terms to be avoided. The Green Book contains definitions, with explanation, of the term 'client' and 'successive owner' and, perhaps most importantly, of 'producer' – 'any party who is liable to the client for the quality of the building as delivered.'

THE MODEL SYSTEM

The Green Book had shown the differences between the major systems described in the National Overviews. W87 has formed the conclusion from these and other studies that no two systems are alike. The example might be taken of residential construction, which in countries such as Belgium, Denmark, Sweden and the United Kingdom, (amongst others) treated by separate legislation, primarily on the grounds that residential occupiers are likely to need greater protection than their counterparts in the commercial sector. In Italy, by contrast, it was observed that no such distinction existed (Green Book p. 181).

These disparities were and, to some extent, still are regarded as likely to inhibit both cross-border construction activity, and the operations of consultants and insurers in other jurisdictions. This was the reasoning behind the EC's interest in the possibility of a degree of harmonisation of post-construction liability arrangements between Member States, of which the reports of Claude Mathurin (a W87 member at that time) were well-known stages.

Nevertheless, the solution could not be seen simply in terms of finding the 'right' national system and encouraging its adoption. Not all systems are equally good or bad; some embody sophisticated approaches to the difficulties encountered, while others are almost incapable of addressing them. In spite of this, no country could claim to have 'all the answers.'

In the light of these overall conclusions, at the 10th Plenary Session at the National Institute of Standards and Technology in Maryland, USA, in October 1994, the Commission invited three of its most distinguished senior members to produce proposals for an 'ideal' or 'model' system. Jens Knocke, the founder Co-ordinator, Professor Gérard Blachère, the former director of the Centre Scientifique et Technique du Bâtiment, and David Barclay, formerly Practice Director at the Royal Institute of British Architects worked together during 1994 and 1995 and consulted with other W87 members.

In February 1996, Jens Knocke, in collaboration with his two colleagues, the triumvirate known affectionately within the Commission as the *éminences grises*, produced CIB Publication 192 'A Model Post-Construction Liability and Insurance System' (ISBN 909-803022-1-X). It sets out proposals for a model post-construction liability and insurance system, with the intention that it should be used not only for academic purposes but to inform the processes of review and reform which have been and are being found necessary in a number of jurisdictions; Australia, Spain and the UK are examples.

The completion of this second W87 publication enabled attention to be fully focused on this, the third project, Case Studies in Post-Construction Liability and Insurance.

Background

W87's preference for publishing the fruits of its work was referred to in Chapter 1. It was the wish of Anthony Lavers, as the new Co-ordinator succeeding Jens Knocke in 1994, to continue this practice. In Tokyo in 1995, in particular, the whole issue of a further major project to be published in due course, was discussed.

From the beginning, the Case Studies were conceived of as a successor volume to the Green Book. It was felt that a further publication would be appropriate for three main reasons. First, and most important, while the National Overviews had provided outline accounts of the post-construction liability and insurance regimes of the respective countries, it would be valuable to provide illustrations of the actual functioning of the systems, showing how situations might be handled. Second, the profile of countries with members of the Commission had changed to some extent, offering the opportunity to provide insights into other systems. Third, a degree of revision might be appropriate in some countries where changes had occurred and where the members could wish to make further comment.

Collaboration

From as long ago as 1993, W87 had identified the possibility of a collaboration with another CIB group. TG15, as it was then known, was a Task Group set up under the joint Co-ordinatorship of Peter Fenn of the University of Manchester Institute of Science and Technology and Edward Davies of Masons, Solicitors. This group, which in 1998 became Working Commission 103, was set up to study Construction Conflict Management and Dispute Resolution. The potential overlap was not seen as a threat to either group but rather as a fertile area for collaboration, and in 1995 and 1996 the then TG15 held its Annual Plenary Session in conjunction with that of W87. The desirability of co-operation in the production of the Case Studies was seen by both groups as almost self-evident. To demonstrate how post-construction liability and insurance systems operated

Case Studies in Post-Construction Liability and Insurance, edited by Anthony Lavers. Published in 1999 by E & FN Spon, 11 New Fetter Lane, London EC4P 4EE, UK. ISBN: 0 419 24570 7

meant providing examples of claims and disputes being dealt with and, in one way or another, resolved. The legal rules and the insurance of the producers' liability was always the province of W87, but every Case Study was to consider outcome and this would necessarily involve negotiation, litigation, arbitration and alternative forms of dispute resolution. In addition, TG15 had members with expertise in countries not available to W87, such as Kuwait and the People's Republic of China. The decision was thus taken to combine resources to undertake the Case Studies Project.

Methodology

Given that a number of the members of W87 and TG15 (W103) are academics or experienced researchers from other backgrounds, it is not surprising that the methodological approach was the source of much anxious and at times heated debate at the joint Plenary Sessions. From the range of possibilities considered, two strands emerged, and these were to become reinforced and entrenched in the discussions.

On the one hand, it was felt that the Case Studies ought to offer the greatest possible scope for comparison and contrast between the different systems and should therefore be based upon a single case. The only way of achieving this would be to use a common basis for the questions to be put to respondents.

On the other hand, some members felt that no hypothetical model could ever replace the reality of genuine cases, and that the Case Studies should be drawn from the extensive experiences, or access to experience, of the members of the respective groups.

This conflict, for thus it manifested itself in the discussions at Plenary Sessions, led to the decision to proceed in both ways. This resulted in the birth of the Type A and Type B Case Studies.

The Type A Case Studies

As has been stated, there was a strong section of opinion amongst the members of W87 and indeed of TG15 in favour of a common base for responses. This was not merely a matter of methodological purity but a genuine concern that one of the benefits to readers would arise from the making of comparisons in the way different systems, some very different, treated the same problematic situation. Self-evidently, this could not be done with real-life cases. The chance of finding identical building failure, parties and legal issues in some twenty countries was remote, to put it no more strongly. Accordingly, it was decided that an hypothetical scenario should be developed to be addressed by all countries responding. The development of the hypothetical scenario was the responsibility of the Co-ordinator, in consultation with the Steering Committee.

The scenario was loosely based on the English case of *Hill Samuel Bank Ltd v Frederick Brand Partnership* (1994) Volume 10 Construction Law Journal No 1 p. 72, although the facts were altered and the issues stylised to suit the needs of the project. In that case, the developers engaged consultants, namely architects and engineers for the design of a business/industrial park south of London. The development included building units with cladding panels made from glass-reinforced cement (known as grc). The plaintiff bank subsequently purchased the development from the developers. The bank found great difficulty in the *disposal* of the units because the grc cladding had sustained hairline surface cracks known as 'crazing' which, although of no significance to the structural soundness or use of the buildings, caused unsightly discoloration. The bank had received certain advice and assurances from the developers' consultants at the time of purchase. The bank's action against the architects and engineers failed. They had no contract with the consultants and tried unsuccessfully to rely upon the law of tort. Ultimately, the architect's avoidance of liability was achieved by a narrow margin because it was held that they had been negligent in failing properly to research the qualities of the grc product which they had incorporated into the design. A Building Research Establishment report on experience of such products could have alerted the architects to the risk of poor aesthetic performance.

In the words of His Honour Judge Michael Rich Q.C. hearing the case as an Official Referee (the specialist construction judges of the High Court, known since October 1998 as judges of the Technology and Construction Court) 'it is more likely than not that (the architects) failed to seek out as much relevant information about these panels as a reasonably competent and diligent architect ought, in respect of an innovative material for which there was no British Standard Specification.'

Despite this damaging finding, the court was not satisfied that the loss suffered by the bank was attributable to the designer's negligence.

The details provided on this case are not given here with any sense of supplying a 'right' answer in relation to the Type A submissions. This would not even be true for the jurisdiction of England and Wales from which it comes. As is explained below, the Type A scenario involved a number of significant changes from the facts of the *Hill Samuel* case, in order to introduce issues of great interest which would have been absent. The reason for giving the real background to the scenario is that the case involved several key points: In particular, it raised the possibility of designers' liability to a purchaser from their developer client, the standard to be expected of designers in the light of published knowledge of state of the art products, the position of developers, and a purchaser's ability to obtain compensation for economic loss arising from poor aesthetic appearance.

The scenario actually distributed to potential respondents was as follows:

> A developer, a company with experience of many commercial developments, commissioned an architect and a contractor for the construction of a 28-unit business park.
>
> The developer did not have any construction professionals on its staff, but some of its senior personnel held strong views on the desired aesthetic appearance of the building. They required the architect to use a particular glass-fibre reinforced concrete product for the cladding tiles.
>
> The architect protested that he had no knowledge of such products, which were manufactured in a foreign country, but the developer insisted. The national building research institute in the country where the building was erected had produced a paper suggesting that such a product might be unsuitable for external cladding because of its tendency to develop superficial hair-line cracks when exposed to weathering. This information was not widely circulated in the construction or architectural journals until about three years after this building was designed. Just under two years after the building was completed, the developer sold it to an investment bank. No inspection of the building was made by the bank.
>
> The bank found that it was impossible to let the units because the superficial cracking on the tiles gave the building a poor aesthetic appearance, although it did not affect the stability or other performance of the building. The directors of the bank had to decide how to deal with this situation.

As can be seen, the scenario adopted incorporates a number of additional factors. The developer's contractor, who was not involved in the *Hill Samuel* litigation, was introduced, to enable consideration of the principal non-professional producer. The client was made to impose his views upon the architect to give the opportunity for discussion of the designer's duty to warn and his relationship with the client. The issue of the possibility of intermediate inspection by the bank as purchaser was known to be relevant in a number of systems, so it was expressly stated to have been omitted. The need for the bank to decide how to deal with the situation was intended to open up the discussion of methods of dispute resolution.

The Type A Scenario was circulated to members of W87 and TG15 with knowledge of a wide range of jurisdictions. The prospective respondents were asked to address the following issues in their submissions.

- The purchaser's opportunity for recovery, through litigation or otherwise, from the vendor.
- The purchaser's opportunity for recovery, through litigation or otherwise, from the vendor's contractor.
- The purchaser's opportunity for recovery, through litigation or otherwise, from the vendor's architect.
- The involvement (if any) of insurers in the purchaser's claim for recovery, including the method of resolving the matter.
- The procedures by which the above are resolved, including extra-legal, legal and insurance procedures.
- Any mechanisms existing for utilising the feed-back or experience from the case or sending it to a body or bodies which may make use of it.

The Type B Case Studies

The second, equally strongly held view amongst members of W87 and TG15 was that the Case Studies should convey a realistic impression of how post-construction liability and insurance issues are decided within the respective systems and how claims and disputes are handled. Accordingly, the author/s covering each country was invited to submit details of cases which had actually occurred within its jurisdiction. The invitation was to submit two case studies, which would be known as B–I and B–II, although one study only was acceptable if that was preferred. The majority of countries returned two case studies. It was suggested that the two should, where possible, cover both residential and non-residential situations and many countries did so.

Whereas the Type A Case Studies were tightly centred around the hypothetical scenario provided, so that few points to be addressed needed to be supplied, with the Type B Case Studies the choice of material available to respondents was extremely wide. Up to a point, this was a virtue, in that respondents could choose what they considered most useful in giving a flavour of their own systems, from the material available to them. However, it was necessary to strike a balance between freedom and guidance, so that the respondents would have an accurate idea of what was required and the aspects of the cases which would need to be addressed, to enable comparisons to be made between the systems. The standard format utilised by Jens Knocke in obtaining the National Overviews for the Green Book was a successful tool and was unashamedly followed for this purpose, albeit, obviously, with very different questions.

The questionnaire sent out comprised some 60 questions or items to be addressed insofar as the respondent found them relevant or appropriate for inclusion. It was in no sense obligatory that each should be covered. The points are set out in the Type B Case Studies section itself, but the categories can be summarised as follows:

- A section on Generalities, requesting information about the authors and other contributors.
- Details of the plaintiff or claimant, such as nature or organisation/individual and the sector to which it belonged (public/private/semi-public).
- Details of the works, identifying the nature of the project, the purpose of the building and any other information, such as innovative design or materials and the time frame, if relevant.
- The procurement method, if relevant, which might be traditional, design and build or other variation.
- Details of insurance held by the parties (if any), such as professional indemnity or building insurance.
- Details of the damage or loss suffered, whether personal injury, repair cost or consequential loss.
- Details of the claims process following discovery of damage/loss, including notification of insurers (if any).
- Details of any technical reports made, identifying those giving instructions and paying for them.
- The method of resolution of the matter. which might be negotiation, litigation or some alternative conflict resolution procedure, including the costs and who paid them, the information available to the parties, the time taken and the role of experts (if any).
- The outcome of the case and how, if at all, the result is the subject of 'feedback' to the public, or the construction and/or insurance industries.

Beyond these quite detailed requests for information, no additional constraints were placed upon the selection of the Type B Case Studies, apart from those of access to information within their own systems. While recent cases were generally preferred, it was considered acceptable to supply older case studies if they were 'classics' of their kind and a handful of respondents did so from choice. In many cases, confidentiality had to be respected and the names of the parties had to be removed and in a few instances insignificant details removed or changed if they would have revealed the identity of the case. Naturally, where the case was in the public domain because it had been reported in a journal or law report this did not apply.

The gathering and collation of the information

To try to ensure some coherence in the responses of countries with more than one member or where some contributions from non-members would be inevitable (not every country has members with expertise in construction and law and insurance), individuals were identified for each country who would be asked to ensure that an appropriate submission was made. It was foreseen that there might need to be an element of 'chasing-up' of delayed responses which would more easily be done within a country than from overseas. These individuals were given the title 'focal points', although their status was informal. The device had varying degrees of success, usually dependent upon the individual and workload of the person asked to act in this capacity and the relative ease or difficulty of securing a response in his/her country.

The first responses were received in the latter part of 1996 and the majority of later ones in the first part of 1997. A small number were received as late as the 1997 Plenary Meeting in September and the Co-ordinator had to seek help in plugging certain gaps actually during the editing process in 1998. Eventually, however, contributions were received from nineteen countries: Argentina, Australia, Canada: Quebec, Chile, China (People's Republic), Denmark, England and Wales, France, Germany, Hong Kong, Japan, Kuwait, Norway, Singapore, Spain, Sweden, Switzerland, Uruguay and the United States of America. The considerable variations in length and depth as well as in approach meant that the editing process, already delayed for other reasons, took longer than anticipated. A lesson which was learned from this project was that flexibility in approach has a heavy price to pay in terms of diversity of responses received. Against that, it should be acknowledged that the variations are in many respects desirable, even if complete uniformity would have been methodologically attractive. Some countries, especially the South American ones, have not appeared in W87 materials previously and greater latitude was given to them in providing background information, especially in the Type A responses, than to those countries, typically European, where for example, reference can be made to the Green Book for the background to the national system. Conversely, some responses, especially of the Type B variety were shorter than average, because the authors were contending with difficulties of confidentiality or access to protected information.

The process of the gathering of the information and its collation was lengthy, occupying most of two years, without the addition of the editing stages. In hindsight, this was to be anticipated from the experience of the Green Book. The results, however, are considered to be a rich source of information on the disparate ways in which legal systems operate in the construction and insurance sectors to address post-construction liability issues.

PART TWO

Argentina

Isaac E. Edelstein

INTRODUCTION

1 Background

In Argentina, a federal republic, the civil responsibility of persons is established in the Civil Code (CC), which is valid in all the provinces of the federation. In each province there are, further to the CC, laws that define and regulate the activities of architects, engineers, technicians and contractors.

The CC does not define professional responsibility but does consider liability for construction defects. The CC, written during the second half of the 19th century, does not establish differences between 'producers', and only refers to the contractor (*el constructor*). Nowadays, the term *constructor* is interpreted as implying any professional or trades person legally entitled to carry out construction works, at any stage, be it design, construction or demolition. There is a considerable growth in the number of lawsuits for professional negligence.

CC Art. 1646 (fundamental in this respect) provides:

> Once the works have been received and paid for by the client, the producer will be held responsible if they suffer '*ruina*', whether partial or total, whether this is due to a construction defect, to a defect in the soil or to bad quality of the materials, whether the producer has provided them or not, and whether built on land provided by the client or not.

This rule extends to soil studies even if these have been provided by the client, as well as to the selection of materials, quality control, design and

Case Studies in Post-Construction Liability and Insurance, edited by Anthony Lavers. Published in 1999 by E & FN Spon, 11 New Fetter Lane, London EC4P 4EE, UK. ISBN: 0 419 24570 7

technical management of the works. Thus, the responsibility of the producer (*el constructor*) is that of a 'duty of result'.

CC Art. 512 furthermore establishes that:

> The *culpa* of the person from whom something is expected encompasses the lack of the diligence required because of the nature of the obligation, and which corresponds to the circumstances of the persons, of the time and the place.

2 The Difference Between Legal Interpretation and Reality

Legal interpretation seems simple and direct enough: the liability of producers is not in doubt, and the term *constructor* renders all who intervene at any stage of the works liable for their activities. Thus, designers, structural engineers, consultants, technical directors, inspectors, supervisors, certifiers, developers, can all be sued and charged individually or *in solidum*, jointly or severally for defects causing the partial or total ruin of the works. In practice, the plaintiff turns against those who are solvent and against whom documented proof exists.

On the other hand, practice calls for some observations. Firstly, producers display a certain ignorance of the responsibility established by the CC: they may be aware of the legal requirements but ignorant of the magnitude of the risks involved; of course, ignorance does not exculpate, and it is felt that the issue needs more emphasis in higher education. Secondly, many works, of varying importance, are executed without intervention of professionals or tradesmen. It is not infrequent that, due to municipal requirements, a professional is *a posteriori* called in to monitor or inspect, execute tests or produce as-built drawings. Professionals may thus find themselves liable for someone else's work.

3 The Liability Situation

The period of the producers' legal liability is ten years. This is stated in CC Art. 1646:

> For the liability to be enforceable, the ruin must take place within ten years of the date of hand-over and the term to initiate action is one year from the moment this occurred.
>
> The liability imposed by this article extends without distinction to the manager of the works and to the designer [though the article's text contains no list, this is,

in practice, a vast array of professionals and trades: *director técnico, representante técnico, inspector, supervisor, certificador, coordinador de las obras* and others], according to the circumstances, without precluding any rights of recourse. Exclusion of liability in contract is not admissible in case of total or partial ruin.

The decennial period starts with the provisional hand-over or with the client's taking possession of the works, whichever is the earlier.

Hand-over in two stages (provisional and final) is prescribed in the Law of Public Works, but not in the CC. Final hand-over can take place up to one year after the works have been taken over by the client; the intervening period is used to correct defects appearing after the provisional hand-over. During this period, it is common for the producer to buy a bond to liberate the retention moneys.

According to CC Art. 1647 bis, the client has 60 days after discovery to report the manifestation of defects hidden at the time of taking possession.

After hand-over, the participation of producers in the use, control and maintenance of the works is minimal. There are no legal requirements concerning the client's maintenance of the works or mitigation of loss or deterioration, except municipal regulations protecting the safety of the occupants, neighbours and the public. Likewise, there are no requirements concerning insurance to cover the decennial liability. This state of affairs may be said to place the producers in an unfavourable position in relation to regulation and common law liability.

Also, the CC does not distinguish between elements with a long life span and those with a shorter one, but according to current case law there is, further to the decennial liability period, a biennial liability known as an 'extra-contractual' liability for unspecified matters such as fungus, dust, noise and others.

4 Limited Extent of Liability Insurance

Recently, insurance companies have been offering liability and professional indemnity cover, albeit hesitantly and on an annual basis which makes renewal dubious. The cover is limited to ARS 50,000 (the cost of an eighty square metre dwelling), and premiums are high.

Professional associations recommend their members to take out liability insurance but do not have the economic power to create mutual insurance companies in accordance with the requirements in the law on insurance. Insurance brokers also point out that re-insurance is difficult to obtain.

Type A

THE PURCHASER'S OPPORTUNITY FOR RECOVERY, THROUGH LITIGATION OR OTHERWISE, FROM THE VENDOR

The purchaser can initiate legal action against a vendor who sells in good faith, although the Bank acquired the property as it was. We assume that there was no inspection of the building, so that the defects were unknown to the vendor as well as to the Bank.

The damage and losses which motivated the claim would be based on a reduction in the value of the property, the difficulty of renting out the premises, the loss in rental income, the cost of repairs, the damage to third parties and 'moral damage' caused by the difficulties and distress suffered. (This *daño moral* usually involves arguments such as damage to reputation, loss of opportunities and hardship.) Inspection by a purchaser is not obligatory, so the Bank cannot be considered negligent. If, on the other hand, it can be proved that the vendor knew about the defect and acted in bad faith, a lawsuit could result in the cancellation of the purchase with corresponding compensation for damage and distress to the purchaser.

In all events, it is also possible to initiate negotiations for an extra-judicial settlement.

THE PURCHASER'S OPPORTUNITY FOR RECOVERY, THROUGH LITIGATION OR OTHERWISE, FROM THE VENDOR'S CONTRACTOR

Legal action against the contractor to recover losses could also be taken if the contractor did not use reliable personnel for the execution of the works, because the contractor would be violating other norms of professional and trade responsibility: the Civil Code (CC) Art. 1631 specifies that 'The contractor is responsible for the works executed by the personnel employed for the works.'

THE PURCHASER'S OPPORTUNITY FOR RECOVERY, THROUGH LITIGATION OR OTHERWISE, FROM THE VENDOR'S ARCHITECT

Legal action could also be initiated against the architect responsible for the design of the works.

THE INVOLVEMENT (IF ANY) OF INSURERS IN THE PURCHASER'S CLAIM FOR RECOVERY, INCLUDING THE METHOD OF RESOLVING THE MATTER

There would be no insurers involved since it is neither obligatory nor customary to obtain insurance to cover liability.

THE PROCEDURES BY WHICH THE ABOVE ARE RESOLVED, INCLUDING EXTRA-LEGAL, LEGAL AND INSURANCE PROCEDURES

The stages in general would be the following:
(a) Preparation of proof, with a technical report describing damage and loss, with an estimate of the cost of repair.
(b) An affidavit before a public notary affirming the essential points of the report and carried out on the site of the damage.
(c) Communication to the producers presumed responsible (the architect and the contractor; *prima facie* both may be liable), with concrete claims for damage and distress, with indication of place, amount, terms for payment and other details.

The judicial proceedings start with the presentation, by the plaintiff's legal counsel, of the case to a civil court with a written description of the facts, other documentation and the claim, in order to notify the defendants. The claim can be made against any of the parties involved in the works, from design to completion, without precluding the defendant from turning against other parties who could be deemed entirely or partially responsible (co-defendants).

If they opt for extra-judicial procedures, the parties would attend meetings for information and clarification, with, as the case may be, the presence of technical and legal experts and arbiters.

During the proceedings, official experts would be designated to advise the court; the parties could also call on their own experts who may disagree with the official expert. Once the evidence has been presented, the parties would present their pleas for a verdict at first instance. It is possible and not unusual to appeal.

ANY MECHANISMS EXISTING FOR UTILISING THE FEEDBACK OR EXPERIENCE FROM THE CASE TO A BODY OR BODIES WHICH MAY MAKE USE OF IT

Since post-construction liability insurance is practically non-existent, we lack feedback from this sector. Summaries of verdicts and jurisprudence are published without any special section for post-construction liability.

There are, to our knowledge, no organisations or publications specialising in claims for civil liability of producers or featuring descriptions of errors in design and construction.

Type B–I

The *Titles* case (fictitious name), relevant to Córdoba

1 The Plaintiff
The plaintiff was a residential condominium (*propiedad horizontal*) in the private sector.

2 The Works
A twelve storey multi-dwelling unit, with commercial premises on the ground floor, located in the city centre, was built with ceramic bricks with a reinforced concrete structure.

3 Procurement Methods
The design was commissioned from a licensed professional, who was also appointed for the technical management, *dirección técnica*. The construction was carried out, on behalf of the future Owners' Association, through a system known as 'by administration'. Under this system, the Associa-

tion's superintendent contracted for the provision of materials and for construction services, and was in charge of payments.

The façade was designed and executed in big panels clad with decorated glazed ceramic tiles. The contract for this particular part of the works included materials and labour.

4 Insurance
The only insurance taken out was against site accidents and fire.

5 Damage or Loss
After hand-over, individual tiles as well as whole parts of the cladding came unstuck.

The damage was discovered by the manager of the property when tiles started falling off.

The cause was established through an expert study on the fallen material and the specifications of the project. The error was in the design, by omission: the material, the base for the cladding and adhesives were not specified. Likewise, there was negligence in the supervision and direction of the works.

10 Process after Discovery
The superintendent of the condominium initiated a claim against the contractor for breach of contract. It should be noted that the architect had not prescribed the technique for fixing the cladding; the Condominium therefore sued the contractor only.

The contractor, assuming that this part of the work was her responsibility, did not sue the architect.

The plaintiff demanded compensation for the cost of
(a) removal of and making good the base for the cladding;
(b) re-cladding of the façade with materials of equivalent value;
(c) losses due to decrease in value of the property;
(d) compensation for non-fulfilment of contract;
(e) compensation for moral damage (*daño moral*: distress, loss of prestige, nuisance...), damage to the public pavement at ground level, claims from neighbours, risks to pedestrians and other claims;
(f) indirect costs due to costly insurance against accidents, rental of equipment and protective scaffolding, further costs for technical management, legal fees, costs of guarding the premises, and other costs;
(g) interest on capital for repair.

The contractor, as mentioned, did not seek recourse against the architect.

11 Technical Reports

The plaintiff commissioned a technical report from an independent expert; the report established both the possible causes and the cost of repair.

The defendants' lawyers also commissioned a technical report for the court hearing.

20 Resolution

The issue was settled by an agreement between the parties, during the judicial process, before the ruling. The contractor offered to pay part of the claim.

21 Costs

Further to the payments outlined above, each party paid its own legal and expert fees.

22 Information

Both parties had access to all studies and reports.

23 Time

The periods of time were approximately
- preparation of claims and substantiating documents: 90 days;
- the defendant's reply to the plaintiff: 30 days;
- the plaintiff's technical report: 40 days;
- initiation of the judicial claim: 30 days;
- nomination of the court's and other experts: 20 days;
- experts' reports: 30 days;
- the defendant's further legal preparation: 15 days;
- the plaintiff's further preparation of argumentation and claim: 30 days;
- submission to the defendant and her preparation of her case: 15 days.

24 Role of Experts

The Court appointed an expert to provide information on a questionnaire submitted by the plaintiff. This expert was nominated at random from a list of registered experts. As mentioned, each party nominated its own experts to act as auditors of the official expert's report.

25 Outcome

The plaintiff agreed to pay fifty per cent of the cost of repair while the contractor met the cost of the materials installed initially, plus fifty percent of the repairs to the base and the re-laying of the cladding, in addition to the fees for the court's expert (incidentally, this, in conjunction with the severe monetary inflation of the time, caused her bankruptcy).

30 Feedback
The details of the case are available neither to the public in general nor to practising professionals or professional associations.

Type B–II

The *Ventanas* case (fictitious name), relevant to Buenos Aires

1 The Plaintiff
The plaintiff was a client frequently commissioning construction works, namely a typical housing institution developing residential units for its members. Its capital was mixed (public and private) as the institution provided social services.

2 The Works
The works were a series of multi-storey buildings with multi-family dwellings for owners subsidised by the government, with long-term financing and located in the residential zone.

The loadbearing structure was of reinforced concrete, with walls of ceramic blocks, metallic windows, concealed pipes; intermediate quality.

3 Procurement Methods
The design was commissioned from a prestigious architects' firm. The technical management was assigned to an associated group of professionals of recognised competence. The works were executed by an experienced local general contractor.

4 Insurance
The only insurance taken out was against site accidents and fire.

5 Damage or Loss

The affected parts were the windows. The damage was mainly due to a fault in the weather strips, which were of synthetic rubber foam; this appeared seven years after the completion of the works, and was blamed on design errors, faulty materials and omissions in supervision.

The damage was observed by the owners who complained to the management.

10 Process after Discovery

The plaintiff initiated a claim against all producers involved in the design, technical management and construction for compensation for

- repair;
- replacement of whole windows as need be; and
- indemnification for 'moral damage' (*daño moral*: distress, loss of prestige, nuisance, disturbance of enjoyment), and loss in the property's value.

As mentioned, none of the producers carried post-construction liability insurance.

11 Technical Reports

All parties commissioned technical reports from independent experts.

20 Resolution

The claim was resolved through an agreement between the parties.

21 Costs

Each party paid its own legal and expert fees. The contractor met part of the cost of repair (see 25).

22 Information

All parties had access to all analyses and studies.

23 Time

The periods of time, in total, some fourteen months, were approximately as follows:

- notification of problems: 80 days;
- administrative proceedings: 120 days;
- technical studies and experts' reports: 90 days;
- attempts to resolve the issue, proposals and agreement ratified by the court: 60 days

24 Role of the Experts
As mentioned, each party nominated its own experts. The court nominated an expert to report.

25 Outcome
According to the technical expertise, the main cause of the damage was the omission of maintenance or lack of replacement of materials with limited life span under varying weather conditions. The issue therefore was settled through an agreement among the parties, as mentioned.

The court ratified the agreement which, *inter alia*, stated that each of the parties pay the fees of their own legal and technical experts. The contractor agreed to pay for the main repairs to the windows, replacement of worn-out elements, reinstating weather strips, replacing rotating elements and paint, as need be.

30 Feedback
The details of the case are not available either to the public in general or to practising professionals or professional associations.

Australia (Victoria)

Type A

Stephen Smith

INTRODUCTION

Australia is a country of six states and two territories. Each state and territory has its own independent legislative building control regime. The case study provides a consideration of the situation in the State of Victoria.

THE PURCHASER'S OPPORTUNITY FOR RECOVERY, THROUGH LITIGATION OR OTHERWISE, FROM THE VENDOR

The nature of the building is a significant factor in the analysis not just of this question, but of the overall rights of the Bank. The fact that the building development was a 28-unit commercial business park, as opposed to a domestic (residential) development, dictates that statutory warranties regarding fitness for purpose and merchantable quality – which run with the title for domestic building works – will not apply. Similarly, the insurance cover, if any, will not be as extensive as would be the case if the development was for residential purposes.

The relationship between the developer as vendor and the Bank as purchaser is one in contract for the sale and purchase of the development. It is a conveyancing transaction, and no building works were the subject of any contract between those parties. The maxim *caveat emptor* (let the buyer beware) applies in relation to a situation where there is a building defect. *Caveat emptor* does not mean that the buyer must take a chance, it means that the buyer must take care (*Wallis v Russell* (1902) 2 IR 585). There is

Case Studies in Post-Construction Liability and Insurance, edited by Anthony Lavers. Published in 1999 by E & FN Spon, 11 New Fetter Lane, London EC4P 4EE, UK. ISBN: 0 419 24570 7

no duty on either the purchaser or the vendor to a contract for the sale of land to disclose to the other party material facts of which he or she is aware and which materially affect the value of the subject matter of the sale (*Terrene Ltd. v Nelson* (1937) 3 All ER 739 at 744). Mere silence by the vendor does not constitute fraud unless there is active concealment of a defect.

In the present case, it would have been commercially prudent on the part of the Bank to take reasonable precautions prior to the purchase by inspecting the premises. If the Bank failed to avail itself of an opportunity to inspect the development, it cannot be said that there has been any concealment of the defect or deceit on the part of the vendor which would found a cause of action by the Bank against the vendor in contract, or the tort of deceit or misleading or deceptive conduct under the Trade Practices Act 1974.

The Supreme Court of Victoria (*Kadissi & Anor. v Jankovic* [1987] VR 255) considered a case where a building was cracked due to inadequate footings, which the purchasers failed to see prior to entering into the contract to purchase land and buildings thereon. The purchasers sought to rescind the contract. The court held that, in the absence of fraudulent concealment or misrepresentation or an express agreement, a vendor of real estate is not liable to a purchaser for defects in a building or land, even if the vendor himself had created the defects in the building or was aware of their existence. Accordingly, the purchasers were liable to complete the purchase of the defective building.

Building defects in the cladding do not constitute a defect in title. In the conveyancing transaction to the Bank, there is no implied warranty by the vendor that the building is fit for occupation or purpose.

Given that the vendor was not the contractor or architect and had no construction professionals on its staff, it would be difficult to attribute any negligence to the developer in using the aesthetically defective cladding, where only expertise would be able reasonably to detect the potential for the unsuitability of the product. Accordingly, the prospect of the Bank recovering from the developer on the basis of negligence is not maintainable.

Thus, there appears to be no real redress available to the Bank against the developer.

THE PURCHASER'S OPPORTUNITY FOR RECOVERY, THROUGH LITIGATION OR OTHERWISE, FROM THE VENDOR'S CONTRACTOR

The vendor's builder ('the contractor') did not contract with the Bank for the construction of the development and no contractual rights accrue to the Bank against the contractor. There is no implied warranty by the contractor that runs with the land to give the Bank a remedy in contract.

The Bank's only avenue of redress against the contractor is to sue in negligence by commencing litigation.

Negligence is the failure of a person having regard to all the circumstances to act as a reasonable and prudent man or woman would have acted.

In considering whether the contractor has been negligent, it will be necessary to consider the following:

(a) Was the building constructed in a proper workmanlike manner so that the defect in the cladding cannot be attributed to the construction methods of the contractor?

(b) Did the contractor conform to all relevant building practices, codes and standards in the construction of the development and in particular the use and affixing of the cladding material?

(c) Did the contractor construct the building in accordance with the plans and specifications, including the manufacturer's specifications?

In relation to consideration (a) above, unless it can be shown that the contractor used unapproved construction methods for affixing the new glass-fibre reinforced cement product or his standard of workmanship was otherwise not of a workmanlike standard, negligence upon this basis is unlikely to be established.

In relation to consideration (b) above, Section 16 of the *Building Act 1993* requires all building works to be carried out in accordance with that Act and the regulations thereunder. The *Building Regulations* 1994 require that building works comply with the *Building Code of Australia* 1996. The development is a Class 5 building under the Code, and the cladding must perform to the required fire rating, weatherproofing and serviceability requirements. The fact that the cladding of the building is simply aesthetically displeasing is not a matter covered by the Code, the Regulations or the Act. However, there are Australian Standards which are highly regarded in the industry as being benchmarks for materials and their methodology of application to buildings. *Australian Standard* AS2908.2 specifically provides for the material requirements of concrete reinforced sheeting and its application for external cladding. To the extent that there has been non-compliance with the Australian Standard, there would be a reasonable basis to assert that the contractor had been negligent and provide a basis for a cause of action in negligence against the contractor.

The use and method of application of a building material which departs from Australian Standards will be *prima facie* defective unless it can be demonstrated by rational analysis that it conforms to accepted standards in the industry (*Bevan Investments Ltd. v Blackhall and Struthers (No. 2)* [1973] 2 NZLR 45).

In relation to paragraph (c) above, if there was any failure on the part of the contractor to provide the cladding in accordance with what was specified in the plans and specifications that may be said to have caused or contributed to the hairline fractures, again a basis for an action in negligence against the contractor may be made out. It is incumbent upon a contractor to use a building product – such as the glass-reinforced concrete product – in accordance with the manufacturer's instructions. In that regard, if the contractor was dealing with a product with which he was unfamiliar, he ought reasonably to contact the manufacturer and seek advice on the suitability of the product for the particular application in which it was intended to be used and heed any advice in that regard and generally in relation to the cladding methodology.

In *Bryan v Maloney* (1995) 128 ALR 163, the High Court of Australia decided that an action in negligence can be brought by a subsequent owner of a building against a negligent contractor (or architect) to recover economic loss arising from a latent defect, and that the date for bringing such action accrues from the date such defect became manifest. At the moment the defect becomes apparent, the value of the development is reduced and economic loss is incurred. From that date, the party who suffers the loss (the Bank) has six years to initiate legal proceedings, with a ten-year limit from the date of issue of the certificate of occupancy being the overriding limitation period.

The cracking in the cladding constitutes a 'latent defect', as it was not apparent until after the building works had been completed.

In an action for negligence, there must be shown to be factual causation between the conduct of the contractor and the defect complained of by the owner. The High Court in *March v Stramare* (1991) 171 CLR 506 decided that the 'but for' test was not determinative of whether there is factual causation in Australia, i.e. that it is not necessary to show that 'but for' the conduct of the contractor the hairline cracking in the building would not have occurred, to show factual causation between the contractor causing or contributing to the hairline cracking. The High Court held that causation must be determined by applying common sense to the facts of each particular case. If it can be shown that the contractor was at least one per cent responsible for the cause of the hairline cracking, he will be responsible for the entire amount of loss and damage, subject to the somewhat presently unclear law of proportionate liability.

In relation to recovery, it is significant to note that, pursuant to the relevant Ministerial Order No. 52 dated 16 May 1996, made under the *Building Act* 1993, contractors are not required to have insurance beyond coverage of AUD 1,000,000 for structural defects. Accordingly, legal action against the contractor for cosmetic defects will not ordinarily be within the scope of the contractor's insurance policy.

The situation is somewhat different in the State of South Australia, which has separate legislation from Victoria. In South Australia the legislative provision specifies that proportionate liability shall be determined by the court not as between 'defendants' but as between the 'persons' responsible for the loss and damage in proportion to their contribution towards such loss and damage.

Bryan v Maloney (1995) 128 ALR 163 also establishes that a contractual exclusion for defects between the contracting parties – being the developer, the contractor and the architect – cannot extend to exclude liability towards a stranger to the contract i.e. the Bank.

In a 1997-case in the County Court of Victoria – *P.V Imports v Minesco Industries* (unreported) –, Judge Hanlon dealt with a case where there was a reinforced glass-fibre cement clad façade of a reception centre which had, as at the date of trial, stood for seven years. The cladding was not in accordance with the manufacturer's specifications and cracks had developed after approximately four years. The defect was primarily aesthetic. The owner sued the contractor and was successful in contract and negligence. However, the owner claimed a replacement façade at a cost of many tens of thousands of dollars. The contractor led evidence that the cracks could be filled with sealant and repainting and the aesthetic appearance of the façade restored for a few thousand dollars. The judge held that the owner was not entitled to a new façade, given the limited aesthetic nature of the defect, the time that had elapsed and the fact that the owner had negotiated the price and a standard of work which did not amount to a 'Rolls Royce' level of finish in the façade. In those circumstances, the owner was not entitled to a 'Rolls Royce' rectification.

THE PURCHASER'S OPPORTUNITY FOR RECOVERY, THROUGH LITIGATION OR OTHERWISE, FROM THE VENDOR'S ARCHITECT

There is no contract between the Bank and the architect, so no action in contract can accrue to the Bank against the architect.

An architect is a professional, and carelessness or unskilful conduct in the design or construction of a building will render him liable to the Bank in negligence for damages for economic loss, measured as either the cost of repairs or the amount of the reduction in value of the development by reason of the defective cladding.

The principle established in *Bryan v Maloney* referred to above has identical application to an architect for the latent defect in the cladding.

It is an important and relevant factor that the architect stated to the developer that he did not have any prior knowledge of the cladding. Notwithstanding the client's insistence that the architect use the particular glass-fibre concrete product for the cladding, the architect will be deemed to have accepted the risk of using that material and being responsible for its failure if there was a reasonable basis upon which he could have avoided the occurrence of the damage. No exclusion clause in the contract with the developer will absolve the architect from liability to the Bank in negligence.

In the circumstances, it would be incumbent upon an architect who has no familiarity with a foreign building product to make enquiries of relevant building associations or bodies as to the suitability of a product for the purposes of a particular construction application. It would not be enough to exclude liability of the architect to say that there were no published articles in construction journals regarding the use or risks in the use of the foreign manufactured cladding product. In that regard, there are strong grounds to support the contention that the architect has been negligent, and if sued by the Bank and found to be at least one per cent negligent, he will be liable for the entire loss and damage established by the Bank to the satisfaction of the court.

In designing the development, the architect must take responsibility for the deployment of a building product in the development which results in latent defects.

The role of the architect is somewhat more technical than that of the contractor, and drawing the plans and specifications for the development makes it the responsibility of the architect to carry out research or engage more specialised expertise, to ensure that the unknown building product is going to be fit for the intended purpose and will yield the desired outcome, both aesthetically and in substance. By his failure to engage a consulting engineer or apparently take any other step to identify the aesthetic and physical properties of the glass-fibre cladding, he runs the risk that he will be found liable in negligence.

The design of a building with the application of building materials which depart from Australian Standards will be *prima facie* defective unless it can be demonstrated by rational analysis that it conforms to accepted standards in the industry (*Bevan Investments Ltd. v Blackhall and Struthers (No. 2)* [1973] 2 NZLR 45). As mentioned above (vis-à-vis the contractor), to the extent that the cladding fails to conform to the requirements of any Australian Standard, a basis for asserting that the architect has been negligent by having due regard to the Australian Standard will be actionable.

In relation to recovery it is significant to note that, pursuant to the relevant Ministerial Order No. 52 dated 16 May 1996, under the *Building Act 1993*, architects are not required to have insurance beyond coverage of AUD 1,000,000 for any act, error or omission in the conduct of their business as an architect. Accordingly, legal action against the architect for cosmetic defects will be within the scope of the architect's insurance policy.

THE INVOLVEMENT (IF ANY) OF INSURERS IN THE PURCHASER'S CLAIM FOR RECOVERY, INCLUDING THE METHOD OF RESOLVING THE MATTER

The developer will not ordinarily have any insurance at all, given the engagement by him of a contractor for the construction and an architect for its design.

The contractor will have mandatory insurance for structural defects in the building arising within ten years of the issue of a *Certificate of Occupancy* following completion of the works.

The architect will have mandatory insurance for any defect in the building arising as a result of any act, error or omission by the architect in the conduct of his business as an architect, arising within ten years of the issue of a *Certificate of Occupancy,* following completion of the works.

Ordinarily, and having regard to the common law principle of 'privity of contract', the Bank would not be entitled to make a claim directly against the contractor's or the architect's insurer as it was not a party to the insurance policy (agreement). In the usual case, the Bank would only have the right to sue the contractor and architect directly, and it is for each of the respective building practitioners to make a claim for indemnity from their professional insurers. However, Australian law provides two bases upon which a claim may be brought directly against an insurer.

Firstly Section 48(1) of the *Insurance Contracts Act 1984* states:

> Where a person who is not a party to a contract of general insurance is specified or referred to in the contract, whether by name or otherwise, as a person to whom the insurance cover provided by the contract extends, that person has a right to recover the amount of the person's loss from the insurer in accordance with the contract notwithstanding that the person is not a party to the contract.

Secondly, the High Court of Australia modified the principle of 'privity of contract' when it dealt with a claim on an insurance policy by a prospective beneficiary. It was held that in insurance cases where the benefit of the policy was for a third party, an exception to the 'privity of contract' rule to avoid injustice may apply: *Trident General Insurance Co. Ltd. v McNiece Bros. Pty. Ltd.* (1988) CLR 177.

Accordingly, there is a basis in Australia for the Bank to proceed directly against the insurer where the architect or the contractor cannot be located or has become bankrupt or gone into liquidation. Otherwise, and in the absence of the right to proceed directly against the insurer, the Bank would have first to obtain the leave of the court to proceed against the architect or the contractor where they were bankrupt or in liquidation.

The Ministerial Order directs building practitioners in various classes as to the minimum mandatory insurance requirements. The contractor requires only structural-defects-cover. Structural defects are defined in the Ministerial Order as including defective design, workmanship or materials, which does or would prevent the continued practical use of the building or part thereof. It is hardly arguable that the defective cladding prevents the development from being utilised for its use as business premises, given that the defect is only of a cosmetic nature.

On the other hand, the architect's insurance is much wider: to cover any act, error or omission by the architect. His omission to carry out any research or to engage a specialist engineer to assess the suitability of the cladding comes within the scope of the mandatory insurance cover of the architect. However, this cover is limited to AUD 1,000,000.

THE PROCEDURES BY WHICH THE ABOVE ARE RESOLVED, INCLUDING EXTRA-LEGAL, LEGAL AND INSURANCE PROCEDURES

The first step would be to have an assessment made of the quantum of the Bank's loss and damage, represented by either the cost of replacing or repairing the cladding or alternatively the diminution in value of the development by reason of the defect.

The next step would be to serve a letter of demand upon the contractor or the architect or both.

Section 131 of the *Building Act* 1993 has attempted to abolish the rule of joint and several liability. However, the Supreme Court of Victoria has in two recent cases – presently the subject of appeal – interpreted the provision to mean that proportionate liability will only apply as between 'defendants' to a proceeding. Accordingly, if the owner was to sue only the contractor, then no matter what the responsibility of the architect may be, if he is not a party to the proceeding, the contractor will be liable to the owner for 100 per cent of the owner's loss if the contractor was adjudged by the Court to be 1 per cent or more liable for the hairline cracking to the cladding.

Given the present state of the law and the prospects of success against the architect, which are perceived to be better than those against the contractor, if the architect failed to respond favourably to the letter of demand, legal proceedings would be issued in court against the architect alone. The relevant court would depend upon the quantum of the Bank's claim. If the present state of the law changed, the contractor could be joined as an additional defendant during the course of, or following, the legal proceedings against the architect.

The courts have power to order the parties to attend a mediation hearing, and this is a common occurrence in the course of building dispute litigation. The mediator cannot force any party to settle the dispute, but he can independently go through the issues with the parties and explore the scope for each party to compromise.

Another and increasingly used mechanism in the course of building dispute litigation is to refer some or all of the issues which are the subject of the legal proceedings to a special referee appointed by the court. The special referee is usually a person with specialised building experience and skills to be able to deal with technical and building industry standards and issues. The referee's determination is ordinarily binding upon the parties, unless fraud can be shown in the referee's determination. The determination is then adopted as an order of the court.

Apart from legal proceedings in court, unless all parties agree in writing, there can be no arbitration of the dispute conducted by the Institute of Arbitrators and Mediators, or any other body or persons.

THE MECHANISMS EXISTING FOR UTILISING THE FEEDBACK OR EXPERIENCE FROM THE CASE TO A BODY OR BODIES WHICH MAY MAKE USE OF IT

Legal proceedings in the Supreme Court are reported, and unreported judgments also occur in the Supreme and County Courts. All such judgments regarding a case such as the one under consideration are utilised by construction lawyers, insurers and legislative consultative bodies.

At the same time, there are bodies, such as the Building Disputes Practitioner Society, which hold regular seminars and publish important cases and developments in the construction law arena. This case would be illustrative of the problems encountered in the case of a commercial aesthetic defect.

The Building Codes Board, which lays down the technical requirements for building design, construction and materials, would also be a relevant body for giving consideration to the present case experience. It could look at the use of the case to give greater performance indicia in relation to external cladding related to cosmetic fitness. However, it may be difficult to determine what is aesthetically pleasing. Some building products are designed to give an aged appearance to external walls, but the cladding in the present case appeared to crack when exposed to weathering. Possibly, a product manufacturer could be required to specify broadly the product's aesthetic characteristics in use.

Type B

Deepak Bajaj with Cliff Roberts

Type B–I

The *Salvatore v Mureli* Case, relevant to New South Wales

1 The plaintiff
The plaintiff was an individual home owner, for whom this was a one-off commission.

2 The Works
The plaintiff owned a suburban cottage for which he commissioned demolition and rebuilding, using traditional full brick construction.

The works, with a contractual value of AUD 159,000, were completed in March 1993.

3 Procurement Method
The contract was negotiated with a single contractor.

4 Insurance
Not applicable.

5 Damage or Loss
The damage was discovered at the end of the contract period. The defects affected roof tiles, floor tiles, the concrete slab, the structure of the roof and sundry other items.

The errors were multiple: materials, workmanship and supervision.

10 Process after Discovery
The plaintiff, alleging breach of contract by improper building work, claimed AUD 80,000 from the contractor for loss of use and loss of employment.

11 Technical Reports
The plaintiff commissioned, at his own expense, a technical report from an architect.

The defendant, as advised by his solicitor, commissioned a technical report from a building consultant.

20 Resolution
No pre-trial settlement could be reached, so the case proceeded to full commercial arbitration, with both parties represented by legal counsel.

21 Costs
The defendant paid all costs as taxed by the arbitrator.

22 Information
All parties had adequate access to data.

23 Time
The matter was settled in thirteen months.

24 Role of Experts
The parties employed their own experts.

The arbitrator relied upon his own experience in the industry and did not seek further expert advice.

25 Outcome
The award went in the plaintiff's favour, though the damages awarded were reduced from AUD 80,000 to AUD 20,000, with costs assessed at AUD 38,000.

No insurance was involved in the construction matter.

30 Feedback
As arbitration proceedings are not published, the details of the matter are not available to the public.

Type B–II

The *Smith v Party Hire* case, relevant to New South Wales

1 The plaintiff
The plaintiff was the contractor, whose client did not often commission construction works, was not an expert in construction matters and had no

access to construction expertise other than consultants who were commissioned only to a limited extent (below).

2 The Works

The project consisted of a suburban workshop and warehouse of 1500 sq.m, with a pre-cast concrete walling system to reduce masonry and to attempt to keep construction time within twelve weeks.

The works were completed in December 1994, with a contractual value of AUD 250,000.

3 Procurement Method

Procurement was by traditional tendering with selected contractors.

4 Insurance

Not applicable.

5 Damage or Loss

The works were not completed because of faulty design, discovered by the contractor four weeks into the contract period. The design errors and omissions were due to the client's inadequate preparation in commissioning the design.

10 Process

The contractor claimed against the client for loss of profit by wrongful termination.

11 Technical reports

The client commissioned and paid for a report from a building consultant.

20 Resolution

Arbitration proceedings were initiated, but the matter was settled by agreement (Note: commercial arbitration in New South Wales is a form of litigation, and is adversarial in form).

Both parties were represented by legal counsel.

21 Costs

The defendant paid some 80 per cent of the legal and arbitration costs.

22 Information

All parties had adequate access to data and information used in the proceedings.

23 Time

Preparation and arbitration took sixteen months (seven days' hearing or equivalent).

The negotiation leading to the agreement after the interim award took one month.

24 Role of experts

The plaintiff used no expert, gave his own evidence and relied on the arbitrator's expertise.

The arbitrator appointed no expert and used his own experience and site visits.

25 Outcome

The arbitrator's interim award went in favour of the plaintiff, i.e. it held the termination by the client to be wrongful. The subsequent agreement and settlement took into account the likely final outcome.

30 Feedback

As arbitration proceedings are not published, the details of the matter are not available to the public.

Canada, Quebec

Daniel Alain Dagenais

Type A

INTRODUCTION

In the first part, we will examine the purchaser's remedies against the vendor, the contractor and the architect, before analysing, in the second part, the more practical issues involved in the present case. Note that the new *Code civil du Québec* (the CCQ) came into force in January 1994. A number of cases still apply the pre-1994 Civil Code of Lower Canada, and a transitional law deals with cases which may straddle the two regimes. However, for the purposes of our present discussion, we will only examine the relevant provisions of the CCQ. This does not relate to the common-law provinces of Canada.

THE PURCHASER'S OPPORTUNITY FOR RECOVERY THROUGH LITIGATION OR OTHERWISE, FROM THE VENDOR

(a) The Contract of Sale

The vendor is bound to deliver the property and to warrant its quality (CCQ Art. 1716). More specifically, the vendor warrants that the property is free from latent defects which render it unfit for the use for which it was intended, or which so diminish its usefulness that the buyer would not have bought it or paid so high a price if he had been aware of them. The

Case Studies in Post-Construction Liability and Insurance, edited by Anthony Lavers. Published in 1999 by E & FN Spon, 11 New Fetter Lane, London EC4P 4EE, UK. ISBN: 0 419 24570 7

vendor is not bound to warrant against latent defects known to the buyer or against patent defects, (CCQ Art. 1726).

The 'professional vendor' is presumed to have known of the defects (CCQ Art. 1729) and cannot exonerate himself from liability unless he has disclosed the defects of which he was, or could have been, aware, and which affect the quality of the property (CCQ Art. 1733).

There are a number of conditions for the consideration of latent defects in property, namely:

the defect must be serious;

(ii) the defect must be unknown to the purchaser at the time of the sale;

(iii) the defect must be hidden (latent);

(iv) the defect must have existed before the sale;

(v) for the purchaser to be entitled not only to a partial or total restitution of the purchase price, but also to damages, the vendor must have known (or have been presumed to have known) of the defects;

(vi) the purchaser who realises that the property is defective must give the vendor written notice to that effect within a reasonable time.

Some further comments should be made on the three latter conditions.

First, there must be a defect and it must be in existence before the sale, so that the vendor will not be held liable for damage which appears as a result of improper use by the purchaser, or because of normal wear and tear. The purchaser thus has the burden of proving that the defect was in existence before the sale, if no physical damage had actually appeared at that time. There is a presumption that such a defect exists at the time of the sale of the property by a professional vendor if the property deteriorates prematurely; premature detoriation would be decided on the basis of comparison with other property of the same nature (CCQ Art. 1729); CCQ Art. 1728, provides that the purchaser has a right to damages if the vendor was aware of the defect or could not have been unaware of it.

The combination of CCQ Arts 1728 and 1729 gives the purchaser the right to claim, from a professional vendor, all the damages suffered because of the defective condition of the property, in addition to the price of the property.

Secondly, the purchaser must give the vendor a written notice of the defect within a reasonable time after discovering it. In the case of defects which appear gradually, the time begins to run on the day the buyer could be expected to have suspected the seriousness and extent of the defect (CCQ Art. 1739). Although the new Civil Code does not require inspection by an expert, it does adopt the 'reasonable person standard' for discovery of defects by the purchaser.

Action must be taken within three years from the manifestation of the damage (CCQ Arts 2925–2926).

In the present case, it seems that, *prima facie*, the Bank can invoke the existence of latent defects since it would not have purchased the building at the vendor's asking price had it known of the defects in the cladding tiles. The defects were serious, since the superficial cracking made the building difficult to rent at normal market prices. The defects were also unknown to the purchaser at the time of the sale. However, there is some question as to whether they were 'hidden'. Although no cracks were noted by the vendor or the purchaser, the cracks had physically begun to develop before the date of the sale. An experienced surveyor could have foreseen the future deterioration of the building from those first signs.

The Bank purchased the building for the purpose of letting units to commercial tenants, and the building was probably worth a considerable amount of money, but the Bank did not conduct any investigation of the premises. Given the importance of the acquisition, and the fact that the Bank sought to rent out the units, it should have taken some steps to survey the building, or to have it inspected, in order to detect any problems in its structure or appearance. In the present case, it seems difficult to contend how any measures short of an inspection by a building expert could constitute a 'prudent and diligent' examination of the property by the purchaser.

A further important element is to determine whether the Bank had any experience in the purchase and sale of commercial real estate, i.e. if it was a 'professional buyer'. If this were the case, a court would probably find that, in failing to perform a more thorough evaluation of the building, it had not acted 'prudently and diligently'. The defect would have to be considered 'patent' for the purposes of the CCQ and the Bank could therefore not invoke CCQ Art. 1726ff against the vendor.

In order to have a remedy against the vendor, the Bank must, as mentioned, have given written notice of the defect within a reasonable time after discovering it (CCQ Art. 1739(1)). The time for this begins to run from the date the buyer could be expected to have suspected the seriousness of the defect. In the present case, we do not known when and in what circumstances the Bank discovered the existence of defects in the cladding tiles. However, CCQ provides that the vendor cannot invoke the time limit if he was presumed to know of the existence of the defects (CCQ Art. 1739(2)). Such is the case where property deteriorates prematurely and the vendor is a 'professional vendor' (CCQ Art. 1729). In the present case, since the developer had run many such projects in the past, he would have to be considered as a 'professional vendor'. Because of the presumption of knowledge by CCQ Art. 1728 he could not invoke the time limit against the purchaser.

If the Bank could make a successful argument for the existence of latent defects, it could ask for restitution of the price paid or one of the remedies explained under (c) below. Furthermore, if the developer 'could not have been unaware' of the existence of defects – which is likely, since the developer is a 'professional vendor' –, the Bank would have a claim for the loss it had suffered because of the defects in the building, no matter which remedy it seeks (CCQ Art. 1728). The vendor's 'good faith' ignorance of the defect is irrelevant, given the presumption of knowledge of the professional vendor.

(b) The Contract of Enterprise

Under the Civil Code of Lower Canada, a purchaser could not invoke against a developer the articles which provide guarantees against the 'perishing of property' (the property suffering severe damage) due to latent defects. The situation is different under CCQ Art. 2212, which states that a developer who sells works he has built or caused to be built is deemed to be a contractor for the purposes of the chapter on Contracts of Enterprise. Therefore, the discussion under (b) for the second question will be applicable to the developer.

(c) General Regime of Obligations

CCQ Art. 1590 is the general provision on the remedies available to the party to whom an obligation is owed (the '*creditor*' of an obligation). It applies in all cases where such a party has the right to demand that an obligation be performed in full, properly and without delay. If the other party (the '*debtor*') defaults on its obligation without justification, specific performance can be required to obtain the resolution of the contract or the reduction of its own correlative obligations, or take any other measure to enforce its right to the performance of the obligation. The creditor may also ask for damages in the alternative.

In this case, the developer had the obligation to warrant the quality of the property sold (CCQ Art. 1716). He has therefore breached his (contractual) obligation of providing a building free from latent defects.

Before exercising his remedies, the creditor must put the debtor in default, so as to give him a reasonable time to perform his obligations; this notice must be in writing (CCQ Art. 1595). If the debtor does not comply with the request, the creditor may take legal action to enforce his rights.

The creditor may ask the court to order the debtor to perform his obligations (CCQ Art. 1601) by issuing a mandatory injunction. However, permanent injunctions are rarely granted in such cases, since the creditor's loss can be compensated in money. Unless there is some urgency, or the debtor has a unique expertise, it is unlikely that a court would require such a debtor to perform his obligations. The creditor may instead perform the obligation himself or cause it to be performed by another party (CCQ Art. 1602), after having notified the debtor by a judicial or extra-judicial demand (CCQ Art. 1595).

The creditor can ask for rescission of the contract if the debtor's default is sufficiently important (CCQ Art. 1604(1)). Since rescission annuls the contract retroactively, it must be ordered by the court. The parties then are restored to their previous positions: the purchaser returns the property to the vendor, while the vendor remits the price to the purchaser.

The purchaser may instead take a *quanti minoris* action (CCQ Art. 1604(3)). He would then be awarded the difference between the price he actually paid for the property and the price he would have paid had he known of the existence of the defects.

Finally, CCQ specifies that the creditor is entitled to damages for any material injury that is a direct consequence of the debtor's fault (CCQ Art. 1607). Damages compensate for the loss suffered as well as for any loss of profits (CCQ Art. 1611). Note that, in this case, damages would be awarded to the Bank only if the vendor knew (or was presumed to have known) of the existence of the defects (CCQ Arts 1727 and 1728). In practice, courts will award damages, in the construction context, to pay for repairs to the works, and to compensate the purchaser for any other damage which he might have incurred (such as the loss of rental revenue between the delivery of the building and the conclusion of the repair works).

Since the developer has sold a defective property to the Bank, the latter can exercise its legal remedies to obtain compensation. The Bank must first put the developer in default to give him the opportunity to inspect the building and correct the defects. If nothing is then done, the purchaser can proceed with court action.

The Bank can ask the court to force the developer (by injunction) to perform his contractual obligations – i.e. to replace the cladding – since repairs are impossible, given the nature of the defect. As was noted above, it is unlikely that the court would grant an injunction to force the developer to replace the defective tiles in the present case, since the purchaser's loss can be adequately compensated by other means.

The Bank could, after notifying the vendor and giving him a reasonable time to act, retain the services of a building contractor to do the replacement work, and then be reimbursed by the developer. However, the reimbursement would be limited if these repairs added value to the building, i.e. if the value of the replacement cladding was higher than that of the defective tiles: the purchaser cannot be enriched at the vendor's expense. Also, the creditor has the duty to mitigate his loss, as he will be responsible for any deterioration that he could have prevented.

The Bank could take the somewhat extreme measure of asking for rescission of the contract and the restitution of the price paid for the property. However, the purchaser would have to show that the defects in the property were sufficiently important to justify rescission. Even if the defects make it difficult to rent the property, it is unlikely that rescission would be granted because there has been no damage to the structure and the performance of the building (e.g. electricity, plumbing, heating) seems to be satisfactory. Remedies less onerous to the developer (such as damages or a reduction of the price) are available to the purchaser.

The Bank may instead keep the property and ask for a partial reimbursement of the price paid for the building, since it would not have paid so much for the building had it known of the defects in the cladding tiles. The court would then examine the relevant circumstances for a proportional reduction of the Bank's correlative obligations (e.g. loss of future rental revenue, cost of repairs, etc. (CCQ Art. 1604(3))).

If the Bank undertook an action in damages in the present case, it could claim the difference between the rental revenue it would have made had the cladding been of adequate quality and the rental revenue it would have obtained had it lowered the rent sufficiently to meet the market demand. Of course, the Bank could also claim any other damages that were a direct consequence of the vendor's fault, since the developer is a professional vendor (CCQ Art. 1728). The vendor's 'good faith' ignorance of the defects in the present case would, as mentioned, be of no relevance.

THE PURCHASER'S OPPORTUNITY FOR RECOVERY, THROUGH LITIGATION OR OTHERWISE, FROM THE VENDOR'S CONTRACTOR

(a) The Contract of Sale

The CCQ states that a manufacturer is bound to warrant the property in the same manner as the vendor: the manufacturer could be held liable to the purchaser if the property were unfit for the use for which it was intended, or if the defects so diminish its usefulness that the purchaser would not have paid so high a price for that property (CCQ Art. 1726). In the context of immovable property, could the contractor who builds the edifice be considered the 'manufacturer' of that property for the purpose of CCQ Art. 1730? In drafting this article, the Codifiers' intent was clearly to cover movable property and, more specifically, consumer products. This provision has not yet been interpreted in the context of a claim for defects in immovable property.

In the present case, since it is the developer who retained the services of the contractor, there is no direct contractual relationship between the contractor and the purchaser on the basis of the contract of sale. However, if CCQ Art. 1730 can be applied to immovable property, then the Bank could sue the 'manufacturer' (i.e. the contractor who performed the construction of the building) for failing to deliver property which was free from latent defects.

(b) The Contract of Enterprise

The provisions in the Contract of Enterprise can be invoked by the client against the contractor for any defects in the works performed. However, can these articles also be invoked by a subsequent owner, though there is no contractual relationship between the parties in such a case? The CCQ has codified an important principle on the transmission of rights resulting from a sale: CCQ Art. 1442 provides that the rights of the vendor pass to his successor (the purchaser) with the title to the property, if they are ancillary to the property or are directly related to it. This means that the rights which are intimately related to the property – such as warranties as to the quality of the property – are passed on to the new owner upon the sale of the property.

In the present case, the developer's rights against the contractor would be passed on to the Bank, since they are related to the quality of the property. Note also that, as mentioned above, a developer is deemed to be a 'contractor' for the purposes of the CCQ's chapter on the Contract of Enterprise CCQ Art. 2124. CCQ Art. 2113 provides that the client who accepts the works without reservation retains his right to pursue his remedies against the contractor in cases of non-apparent defects or non-apparent poor workmanship: the client's silence on these issues does not constitute a tacit renunciation of his rights against the contractor. One would have to refer to CCQ Art. 1726 in the chapter on sale for the definition of an 'apparent defect': it is a defect that can be perceived by a prudent and diligent buyer, without any need of expert assistance. Of course, similar factual difficulties in determining what may constitute an 'apparent defect' at the time the building was transferred to the client – presumably after the client's experts (his architect or engineer) had recommended its acceptance – will also be present in the interpretation of CCQ Art. 2114.

There are two important provisions under the Contract of Enterprise that may give the client a remedy. First, CCQ Art. 2118 provides that the producers (the contractor, the sub-contractors and the architect or engineer who directed or supervised the work), are jointly and severally liable for the loss of the work occurring within five years after its completion. There must have been damage to an important part of the works, and the loss must have been major: although a total 'perishing' (loss) of the works is not necessary, the damage caused to the building must have made it considerably less functional. The crumbling of walls of chimneys, the loss of insulation and serious leaks in roofing constitute examples of 'losses' envisaged by CCQ Art. 2118.

This warranty applies for five years after the work has been completed. Usually, a certificate indicating the end of the works is issued to the contractor by one of the professionals who has supervision responsibilities with respect to the work (such as an architect). The CCQ states that the work is completed when it has been produced and is ready to be used for its intended purpose (Art. 2110).

In principle, the contractor is under a duty of result: he can only exonerate himself from responsibility for defects in the works if he can prove that they resulted from erroneous or faulty expert opinions or plans supplied by the client's engineer or architect (CCQ Art. 2119(2)). However, the new Code has added additional grounds for the exoneration of the parties involved in a construction project, i.e. if the defect results from decisions imposed by the client in selecting the land, materials, sub-contractors, experts or construction methods (CCQ Art. 2119(3)). The client's 'interference' in the project may therefore exonerate the producers. But the development risks (or 'state of the art risks') usually fall upon the producers, under CCQ Art. 2118 and 2119: the fact that they were unaware of the building research institute's paper, or the existence or non-existence of this paper, is irrelevant.

The client can also claim against the contractor on the basis of the warranty against poor workmanship (CCQ Art. 2120). Producers are held severally liable to warrant the property against poor workmanship for one year after the acceptance of the works. This article protects the purchaser against minor defects, which might arise during a short time after the end of the works; it allows for a certain 'tryout period' and it is unlikely that defects which appear during that period would be caused by normal wear and tear.

By CCQ Art. 1442 the Bank can exercise the vendor's rights under its contracts with the architect and the contractor. If the relevant limitation period has not elapsed, the developer's acceptance of the building does not bar the purchaser from invoking non-apparent defects in the works (CCQ Art. 2113).

Although the hairline cracks gave the building a shabby appearance, they did not affect the stability of the structure or impair its performance. Since only superficial damage is involved, it would be difficult for the purchaser to invoke CCQ Art. 2118. The purchaser could only use such an argument if there was a reasonable probability that tiles would fall off.

However, the Bank has not been able to let the units and the market has proved that the building is unfit for the purpose for which it was built. A strong argument can be made that there has been a 'loss of the works', pursuant to CCQ Art. 2118.

The Bank could raise an argument based on CCQ Art. 2120 – warranty against faulty workmanship – since the damage to the façade was caused by the inadequate quality of the tiles approved by the architect or negligent installation by the contractor. However, this warranty only lasts for one year after the acceptance of the work by the purchaser. In the present case, the developer owned the property for almost two years before selling it to the Bank; the warranty under CCQ Art. 2120 has therefore expired.

In any case, if a court decided that CCQ Art. 2118 would apply to the present case, the contractor could, to exonerate himself, raise the argument that the owner insisted on the choice of the tiles (CCQ Art. 2119(3)). This raises a series of issues, which, for the moment, remain unanswered:

(i) how 'educated' must the client's choice be?

(ii) is the producer bound to inform the client of the defects, especially when they are hidden (CCQ Art. 2104)?

(iii) can the architect avoid any liability by simply saying he was not aware of the defect, or is he bound to assist his client by conducting some investigation?

Note that the purchaser has three years after the first manifestation of the damage to file its claim with the Quebec Superior Court (CCQ Arts 2925–2926).

(c) The General Regime of Obligations

Since the purchaser did not have a direct contractual relationship with the contractor, does he have a claim in extra-contractual (i.e. delictual or tortious) liability against him for the violation of the general norms of reasonable conduct (CCQ Art. 1457)? As we have seen, through the sale of the property, the rights of the vendor, which are closely related to the property, pass to the purchaser (CCQ Art. 1442). The CCQ prevents the purchaser from opting for the rules of extra-contractual liability if it has a contractual relationship with another party. Therefore, in the present case, the Bank would have no choice but to sue the contractor on a contractual basis.

(d) Remedies

The Bank must first put the contractor on written notice (CCQ Art. 1590) that he is in default of his contractual obligations and give him a reasonable time to perform his obligations. If the contractor refuses to do so, the Bank can then exercise its legal remedies.

First, it is unlikely that a court would issue an injunction to force the contractor to replace the tiles, since less coercive remedies are available. On the other hand, the Bank may have the work performed by a third party and sue the contractor for the costs of replacement of the tiles. A *quanti minoris* action is not available in the present case, since the contract between the developer and the contractor is a 'Contract of Enterprise' (a contract to carry out work), not a contract of sale. Finally, the Bank is entitled to all damages that resulted from the contractor's fault. This would notably include the loss of rental revenue between the date of delivery of the defective property and the correction of the defects.

THE PURCHASER'S OPPORTUNITY FOR RECOVERY, THROUGH LITIGATION OR OTHERWISE, FROM VENDOR'S ARCHITECT

(a) The Contract of Enterprise

As was the case for the contractor, the purchaser of a property does not seem to have a contractual remedy against the architect who designed the plans for that property, since there is no contractual relationship between them. However, by virtue of CCQ Art. 1442, the first owner's claim against the architect for his defective design or choice of materials goes with the property into the hands of the new owner. The Bank could therefore make a claim against the developer's architect for the defects in his design or for his defective choice of materials.

The purchaser could invoke the five-year warranty against the 'loss' of the work (CCQ Art. 2118), in his suit against the architect. As was discussed above, CCQ Art. 2118 only applies if a major part of the building is involved and if the loss is substantial.

Second, the purchaser could also invoke the one-year warranty against faulty workmanship (CCQ Art. 2120) which applies to the architect involved in the project, but this warranty has expired because the developer waited two years before selling the building. Note that both CCQ Arts 2118 and 2120 can be invoked against an architect only if he directed or supervised the works.

If the architect did not have any supervision responsibilities, he could be held liable for the losses occasioned by errors in the plans or expert opinions furnished by him (CCQ Art. 2121). The author believes that this article must be understood in the context of CCQ Arts 2118 and 2120 (and their respective limitation periods), and that it does not apply more generally to all cases where damage is caused to someone's property by the architect's fault. If the latter interpretation were adopted, this would mean that the architect could be exposed to liability for an indefinite period of time for his professional errors, since no limitation period is stipulated in CCQ Art. 2121. In the present case, it is not certain whether the architect actually supervised the project or only provided the plans and technical expertise. The courts have yet to interpret CCQ Art. 2121 and its relationship with CCQ Arts 2118 and 2120.

In the CCQ the architect can be relieved from liability if the defects in the work do not result from an error on his part, either in his plans or expert opinions or in his supervision of the works (CCQ Art. 2119(1)). Furthermore, the architect will be exonerated from liability if the defects in the property result from decisions made by the client in selecting the materials, the sub-contractors, the construction methods, etc. (CCQ Art. 2119(3)). These arguments might be particularly useful for the architect in the present case, since he did not commit an error in design, and since the choice of materials was virtually 'imposed' by the developer, subject to the issues raised above.

(b) General Regime of Obligations

As was the case for the contractor, there is no direct contractual relationship between the purchaser of the property and the architect, since it is the developer who concluded separate contracts with each of them. However, because CCQ Art. 1442 operates a transmission of contractual rights from the developer to the Bank, the latter is considered to be in a contractual relationship with the contractor and the architect. Because the CCQ prevents the parties to a contract from opting for other rules of liability (i.e. the extra-contractual regime, CCQ Art. 1458), the purchaser cannot sue the architect on an extra-contractual basis.

The purchaser could argue that the architect negligently performed his professional responsibilities and therefore is liable for a violation of his general contractual obligations (CCQ Art. 1590). However, in the present case, a number of factual elements would have to be considered in deciding this question. First, the architect had expressed some hesitation concerning the choice of tiles, with which he was unfamiliar, but the developer insisted on their use. Some Quebec case law suggests that the architect has a duty to analyse the quality of the materials to be used in construction and to inform the client if his methods were contrary to the 'rules of the art' (standard practice, principles of good practice), especially if the developer had no construction professionals on his staff. Second, the existence of the building research institute's study on the cracking of the tiles as a result of weather exposure suggests that the architect could have discovered the potential defects in the tiles if he had made the necessary inquiries or laboratory tests. Even if no information appeared in construction journals until three years after the design was completed, the hazards inherent in this type of tile were not completely unknown.

However, in the architect's defence, the fact that the information on the tiles' defects had not circulated in any professional journals within three years after the design suggests that he did not act negligently in approving their use at that time. Furthermore, as was mentioned above, the architect can invoke the client's instructions as to the choice of tiles as a cause of exoneration (CCQ Art. 2119(3)). Therefore, a court, in deciding whether the professional should be held liable for the purchaser's losses, would have to balance the developer's insistence on the choice of the tiles with the architect's duty to provide reasonable professional advice as a 'man of the art' (a competent practitioner of the profession).

(c) Remedies

See the discussion under (d) above.

THE INVOLVEMENT (IF ANY) OF INSURERS IN THE PURCHASER'S CLAIM FOR RECOVERY, INCLUDING THE METHOD OF RESOLVING THE MATTER

The purchaser himself would not be insured for the damage in question.

If the purchaser successfully sued the developer, would the latter's liability then be covered by an insurance policy? Probably not, because in most North American jurisdictions, liability insurance policies exclude the cost of making good faulty materials, workmanship or design. The same is true for property owners' insurance. In the present case, since the deterioration of the building's façade was caused by faulty materials or design, the developer would not be covered for the damage caused to the purchaser.

For the same reasons, the contractor has no cover under his liability policy for the replacement of defective goods or faulty workmanship (note that the situation would have been different if a tile had crumbled and injured a passer-by: damage caused to third parties or their property is covered under liability insurance policies). The contractor can take out a performance bond to ensure that his obligations will be executed if he fails to perform, but this bond applies for the duration of the project and may be extended only for a short time afterwards (often one year, which is the usual length of the contractual guarantee of quality).

The architect, for his part, has a professional liability insurance policy. This policy provides cover for damage due to errors and omissions in the performance of his professional responsibilities. It usually covers the professional on a continual basis – rather than by project – and can be renewed on a yearly basis. In the present case, the policy would cover the architect's liability for the damage caused by his approval of the defective tiles. Since it seems that the architect is the only party with the benefit of insurance cover in the present case, a court might be tempted to consider that he should bear the brunt of liability.

THE PROCEDURES BY WHICH THE ABOVE ARE RESOLVED, INCLUDING EXTRA-LEGAL, LEGAL AND INSURANCE PROCEDURES

Since the Bank acquired a commercial building, it could not claim indemnification from the Quebec 'New Homes Warranty Programs', which are available for claims by purchasers of residential property.

While it is unlikely that the contract of sale concluded between the purchaser and the developer would include an arbitration clause, the developer's contract of enterprise with the contractor and his contract for professional services with the architect would probably contain such a clause. The developer could invoke this clause to resolve, perhaps out of court, any disputes with either the contractor or the architect relative to the performance of the building. However, the arbitration clause usually applies to disputes that arise during construction, rather than in the post-construction context.

Unless an out-of-court settlement is reached, the purchaser's claims against the vendor, the contractor and the architect will have to be decided at trial, before the Quebec Superior Court (which has jurisdiction where the object of litigation has a value equal to or greater than CND 30,000). The case proceeds once a 'Certificate of Readiness' has been issued by the Clerk of the Superior Court to indicate that all the requisite procedures and documents have been filed with the Court. In a case like this, our experience shows that it may take between two and four years to file the procedures and documents. The parties are then allocated court time for the hearing, which takes place within six to eighteen months after the Certificate of Readiness has been issued. The Superior Court's final decision may be appealed to the Quebec Court of Appeal, but leave must be obtained from the Supreme Court of Canada, which is the supreme judicial authority for all cases heard in Canada.

Note that at any point of the process prior to the Quebec Superior Court's decision, the parties may choose to use mediation. Note also that if the action is instituted in the district of Montreal, then, once the Certificate of Readiness is issued, they will be invited to use the mediation process organised by the Quebec Bar Association, together with the Superior Court of Montreal. Although this process is voluntary, it is often successful in helping parties find common ground and avoid protracted litigation.

ANY MECHANISMS EXISTING FOR UTILISING THE FEEDBACK OR EXPERIENCE FROM THE CASE TO A BODY OR BODIES WHICH MAY MAKE USE OF IT

None.

Type B–I

Daniel Alain Dagenais with Laurent Arsenault

The *Inuit Civic Centre* case. Relevant to Quebec

1 The Plaintiffs

The plaintiffs were a small and remote mainly Inuit community, represented by an intermediary regional government.

2 The Works

The building was intended as a civic centre for the Community, with recreational facilities and municipal offices.

The facility was a single-storey 24 m x 24 m steel-framed and metal-clad structure. The building's foundations consisted of reinforced concrete monolithic strip footings and floor slabs over polystyrene insulation laid on an engineered gravel pad built on the existing natural permanently frozen soils (permafrost). This so-called 'insulated pad system' represented a departure from the more conventional naturally ventilated crawl-space approach whereby structures are built on stilts above the ground to prevent permafrost deterioration by thawing.

The building was situated in a 95 per cent Inuit village of some 4,000 people in an extremely remote area of Northern Quebec where construction activity is limited to a three and a half months' period between mid-July and late October, as determined by navigation constraints and shipping schedules. Construction thus began in early August and was substantially completed in November 1982, at a cost of approximately CND 1,000,000. Neither provisional nor final acceptance was ever granted, as problems began to appear only 3 months after completion.

3 Procurement Method

A traditional procurement method was used, the Inuit Community being the client. The only exception was that the regional government intervened to control the budget, with the assistance of the Quebec Ministry of Municipal Affairs. Also, it was part of the construction contract that the Community was to provide local labour.

4 Insurance

As usual in Quebec and in the rest of Canada, there was no damage insurance.

As usual also, all design professionals carried professional liability insurance, and the contractor did not.

5 Damage or Loss

As early as three months after completion, cracks started to appear.

The damage was discovered by the owner, who immediately requested the design professionals and the contractor to inspect and propose a solution.

Eventually, the building was condemned. The insulated pad system had proved insufficient: the permafrost was deteriorating by thawing, thus weakening the support of the building.

The plaintiffs instituted an action against the producers, claiming the cost of making good. They were also claiming for the loss of use of the building and for loss of rental income.

10 Process after Discovery

The claim was initiated by the owner, through a notification to the design professionals and the contractor. The designers immediately informed their insurer (they were all insured by the same insurer), who retained the services of experts and attorneys.

11 Technical Reports

The experts retained by the designers' insurer issued a technical report proposing a possible solution. This report was requested and paid for by the insurer and was issued within a few months after a first notice of the loss. A firm tender for remedial work was obtained from a contractor.

The report was transmitted to the plaintiffs, who showed some interest. They were, however, disappointed when they learned that the insurer would not pay for the costs of the repairs at that time.

The plaintiffs then retained the services of independent consultants, sharing the costs of this second report with the regional government. These consultants proposed a different solution.

A meeting was held with all parties and the two groups of experts, but no agreement could be reached. The owner and the regional government then decided to undertake the repair works at their cost, following the method recommended by their own consultants.

This second solution proved to be much more expensive than the first one. It also proved to be more costly in time; the first stage, which required the freezing of the pad, took approximately four years to complete. During that period, court proceedings were stalled. After that delay, the action was amended to reflect the total costs of the first stage of the repair works, with the result that the total amount claimed exceeded the original costs of the building by approximately 40 per cent (CND 400,000).

20 Resolution
At the initiative of the designers' liability insurer, a mediator was retained to organise a sequence of meetings to discuss alternative dispute resolution solutions. A first such meeting was held at the municipality's offices and served to re-establish direct contact between the parties.

21 Costs
At the onset, the mediator's fees were to be shared between the owner and the insurer. It was part of the solution to have them paid only by the insurer.

22 Information
All information needed was available to all parties and could be exchanged either by consent or during examinations on discovery.

23 Time
The conflict resolution process started in November 1993, when the insurer decided to retain the services of the mediator. It took the mediator six months to convince the owner that such a process would be worthwhile and to organise the first meeting. The mediation meetings themselves were conducted over a period of one year; an agreement was reached in April 1995.

24 Role of Experts
Each party used its own experts, except for the contractor who felt that he obviously had no liability, as there were indications that the problem was

related strictly to the design (the contractor would have had to retain an expert however, if the case had been taken to trial).

25 Outcome

The decision to hold the first mediation meeting in the Community was deliberately designed to display a real desire to come forward with a changed attitude towards this dispute. The extreme remoteness of the site and the large number of individuals involved imposed significant expenses on the insurer, but these were deemed necessary to initiate a reversal of perception within the local community, which had understandably felt let down, not just by the professionals but also by all other related parties, all non-native.

The accumulated damages, as well as the interest on the likely exposure of the insurer, created a situation where the economic equivalent for the insurer of its best-case scenario trial outcome was large enough to allow an attractive alternative compensation package to be offered. The Community's needs were discussed, and a tentative agreement was achieved, in principle, on the more significant components of compensation. The insurer thus offered to meet the cost of partial repairs, transforming the abandoned Civic Centre into a municipal garage and warehouse, and the construction of a new administrative building, which the Community sorely needed but could not afford.

Other meetings were needed to fine-tune this package in terms of physical characteristics of the proposed facilities and treatment of outstanding monetary issues. An agreement was reached and the insurer agreed to defer the signing of releases until the construction programme had been completed and received final acceptance, i.e. one year after completion. The facilities were successfully built by October 1995 and have been fully functional since then.

A most significant benefit gained from this conflict resolution approach was that the professionals' reputation and standing with the Community, as well as with its Regional Government, were restored and may lead to future appointments. The insurer allowed this to happen by getting very involved in the execution of a complex compensation package and accepting the costs and inherent risks of doing so, as opposed to simply paying a sum of money in exchange for a release.

30 Feedback

As the mediation was confidential, the case is not available to the public. In Quebec and in Canada, there is none, or very little, feedback available; no feedback system has ever been put in place.

Chile

Corporación: Alberto Ureta A., Juan Carlos León F. and Gastón Escudero with Cristián Canales P. and Juan Claudio Molina

INTRODUCTION

1 General

The issue of post-construction dispute resolution raises several considerations about the legal system: it not only involves specific substantive rules on civil liability – which are those that in the end provide a resolution to each case – but also procedural rules pursuant to which the courts must act; relevant also is the organisation of the courts in the country, and even the culture and custom of private companies in the industrial sector regarding the way they resolve their disputes.

1.1 THE CHILEAN LEGAL SYSTEM

Chilean law belongs to the 'Western Group' of legal systems, being more or less liberal in inspiration, or at times based on a certain state socialism, influenced by Christian morals and a philosophy that emphasises the rights of individuals. Within the group, Chilean law belongs to the 'legal' or Roman-Germanic system.

The main source of law is legislation, although its application lies with the judicial branch, at the head of which is the Supreme Court, whose mission is to safeguard correct enforcement of the law and reverse decisions that violate it.

The laws contained in regulatory texts are called *codes*, such as the Civil Code (CC), the Criminal Code, the Labour Code etc. Codes are complemented by laws aimed at specific matters which are ultimately regulated in a co-ordinated fashion by special and general regulations: Anything not stipulated in the special regulations is bound by the general

Case Studies in Post-Construction Liability and Insurance, edited by Anthony Lavers. Published in 1999 by E & FN Spon, 11 New Fetter Lane, London EC4P 4EE, UK. ISBN: 0 419 24570 7

regulations: special rules have priority over general rules, the latter being enforceable when the former are silent.

In Chile, the CC – governing relations between individuals – was enacted in 1855; it embodies many of the principles of the era, which derived mainly from French and Spanish law.

1.2 SOURCES OF LAW IN CHILE

The first Article of the CC states that a law is 'any statement of sovereign intent that orders, prohibits or permits, when expressed in the manner prescribed by the Constitution.' This definition enunciates the existence of a hierarchy. The Chilean legal system recognises legislation as a principal source of law. The first law, called also the fundamental law, is the Constitution, enacted in 1980. It establishes the political organisation – and sometimes even the economic organisation – of the country, the general principles on which all national legislation must be based, and the procedure by which laws are validly decreed.

The laws decreed by the political authority in accordance with the Constitution are valid in general and applicable to any type of act and to all individuals. Notwithstanding this general nature of the law, individuals may voluntarily bind themselves to liability in law. This principle is established in CC Art.1545: 'Any contract legally entered into is law for the contracting parties.'

Other sources of law of less importance are custom – repeated uniformly and constantly over time and which the members of the society recognise as having an influence on their behaviour –, and general principles of rights, equity, case law and doctrine.

1.2.1 Article four of the Commercial Code states: 'Trade custom replaces the silence of the law' (in civil matters).

1.2.2 The general rules of rights are generally accepted notions regarding how the law must be enforced. They become a source by being embodied in legislation and by having an influence on the criteria used by judges to enforce the law.

1.2.3 Equity is 'the sure and spontaneous feeling of fairness and unfairness that derives from human nature only, dispensing with positive [written] law.'[1] In Chile, equity has a value as a source of law, but in a very restricted way: 'Whenever the preceding rules of interpretation cannot be applied, obscure or contradictory passages shall be interpreted in a way that most seemingly conforms to the general spirit of the legislation and natural equity' (CC Art. 24). Consequently, equity is an element of interpretation that the judge must apply when the rules of interpretation indicated in the CC are not applicable.

Moreover, at times the law refers to equity in situations that, given their complexity or individuality, cannot be reduced to a general rule. Finally, there are cases where there are gaps in the law, meaning that there are no stipulations therein, and therefore decisions must be dictated according to the principles of equity (Code of Civil Procedure Art. 170, paragraph 5).

1.2.4 CC Art. 3 provides that 'judicial decisions have no binding force except regarding the cases in which they are actually rendered.' Therefore, in Chile, decisions are not valid for similar subsequent cases. However, in practice, judges at times consider the way in which a case has been resolved previously, particularly when there is uniform case law on such a matter, and, above all, when the Supreme Court has made a ruling on the issue. We can state that a court is not bound by the decisions which it or another court has adopted previously, even if the latter is a superior court, and the court may change its opinion or sustain contrary criteria. Nevertheless, past decisions constitute a precedent that courts do consider on occasion, especially in the case of decisions of the Supreme Court.

1.2.5 Finally, doctrine may constitute yet another source of law. Doctrinal interpretation is expressed in treatises, law reviews and teachings, and commonly evolves faster than the law, since it closely follows social changes. However, it has no binding force, and in practice serves only as an orientation for legislators and judges, depending on the prestige of the interpreter or commentator.

1.3 PROCEDURES

In this chapter, we will consider some aspects of the different forms by which a dispute may be resolved in our legal system, whether the parties seek to enforce the relevant laws before the courts or choose extra-legal means, such as cases where the parties wish to remove a dispute from the ordinary court in order to submit it to a tribunal appointed either by the parties themselves, or by the judicial authority. Such tribunals are known as arbitrators or arbitration panels.

In addition, we will consider cases where the parties may seek a solution privately, i.e. without resorting to the State judicial system or to an arbitrator-at-law; this is of particular relevance in the area of construction disputes.

1.3.1 Courts

There is a Bill before Parliament that will make it possible to hear construction disputes not in ordinary proceedings, but in an accelerated procedure.

1.3.2 Arbitration Panels

These are comprised of arbitrators-at-law appointed by the parties, by the judicial authority, by law or by the will of the testator for the resolution of matters in dispute.

Arbitrators-at-law are classified as:

- Arbitrators-at-law, who conduct hearings and make decisions like ordinary courts, meaning subject strictly to the law.
- Arbitrators *ex aequo et bono*, or conciliators, who proceed in accordance with the rules of procedure agreed upon by the parties or, alternatively, according to the minimum rules established in procedural law, but whose decisions are based on equity.
- Mixed arbitrators, who proceed strictly subject to law but whose decisions are also equitably based.

Arbitration panels are different from ordinary courts because the latter derive their jurisdiction from delegation of the state's powers, while the former are given power by the parties who appoint them. Moreover, ordinary courts have the 'power of rule' – the power to have the judgment enforced by themselves, resorting to support from public forces as need be –, while arbitrators must resort to ordinary courts to have their rulings enforced.

In regard to the matters that may be submitted to arbitration, procedural law makes the following classification:

(a) Matters for which arbitration is prohibited, i.e. matters that the law prevents from being submitted to the decision of arbitrators for reasons of public convenience. The Organic Code of Courts Arts 229 and 230 mention the following: alimony, separation of estates between a husband and wife, matters in which a public ministry must be heard, and matters arising between a legal representative and the principle thereof.

(b) Matters submitted mandatorily to arbitration, i.e. matters that the law delivers specifically to the decision of arbitrators-at-law. By way of example, the Organic Code of Courts Art. 227 mentions the liquidation of a marital partnership, the division of assets, matters arising because of the presentation of the account of a manager or liquidator of business corporations and of differences arising between the partners in certain business corporations.

(c) Matters submitted voluntarily to arbitration, i.e. matters that the parties may submit to the decision of an arbitrator by mutual consent. They constitute the general rule.

It is worth adding that a Bill is now before Parliament that expands involuntary arbitration to new matters, including controversies that arise in relation to construction acts and contracts when the amount thereof is greater than 2,000 'tax units' (currently some USD 110,000). Below this amount, the plaintiff would have the power to submit the matter to the ordinary courts of justice or to an arbitrator-at-law, at his discretion.

Currently, controversies relative to construction contracts are matters for voluntary arbitration. In practice, parties often stipulate that any dispute that may arise must be resolved by arbitration.

In view of the bureaucracy and trouble implied by resorting to ordinary justice, arbitration is an attractive alternative, particularly when the controversy arises in an industrial or commercial activity where speed and efficiency in the resolution of conflicts are imperious needs. For this reason, the Santiago Chamber of Commerce in 1994 created the *Centre of Arbitration of Santiago*, the purpose of which is to administer national and international arbitration submitted to it and to appoint arbitrators and conciliators when the parties have so agreed.

It may be assumed that the existence of this organisation will foster arbitration justice as an alternative to court resolution of commercial and economic conflicts, including, of course, the construction sector. However, of the 24 cases that have so far been completed by arbitrators from the Centre, none relates to construction. This circumstance tends to reflect the fact that this type of matter is often not heard before judicial or quasi-judicial bodies, nor resolved by litigation.

1.4 NON-CONTENTIOUS MEANS OF DISPUTE RESOLUTION

By the expression 'non-contentious means,' we refer to any non-conflictual or peaceful mechanism for dispute resolution used by the parties. In connection with construction, it is very common for the parties to seek a resolution to their differences privately and amicably, given the cost in terms of time, expense and trouble implied by a lawsuit before an ordinary court.

One way to do so is by means of direct negotiation where an agreement is reached regarding the amount of the damages and the way to pay them. This agreement may be verbal or it may take the legal form of a settlement agreement, as established in CC Art. 2446, and defined as a 'contract where the parties extra-judicially terminate pending litigation or provide against eventual litigation.' The settlement then produces an effect similar to a final ruling.

Another way to resolve a controversy peacefully is to resort to a third party who, in the capacity of mediator, hears the positions of the parties and conciliates their interests in order to reach a final consensus. Unlike arbitration, where the third party defines a solution that the parties must abide by whether satisfied or not, here the role of the third party is not to issue and impose a solution, but rather to facilitate an understanding in order to reach a mutually satisfactory agreement.

2 Construction-specific matters

In these remarks, we shall indicate in general the legal and regulatory rules applicable to the solution of conflicts specifically arising from defects once the construction has been handed over to the party who commissioned their execution (the client). The different types of liabilities that may be attributed to the persons involved in the construction will be described.

These remarks mainly explore the legal and regulatory rules applicable to the Case A study but may also be seen as a general description of the Chilean situation.

2.1 LIABILITY IN THE CHILEAN LEGAL SYSTEM

According to a widely accepted definition, liability is 'the obligation weighing on a person to compensate for damage suffered by another.' [2]

Liability may be based on a violation of a contract – contractual liability –, the commission of a civil offence or quasi-offence – extra-contractual liability –, or the commission of a criminal offence – criminal liability

2.1.1 Contractual liability

This liability arises from the violation of a contract or agreement understood as any agreement made with the intent to create, modify or extinguish rights and obligations.

As a general rule, if one of the parties has defaulted on its obligations under a contract or agreement, the other party, who has not defaulted or is willing to comply, can either request forced compliance with such obligations or termination of the contract, in both cases with compensation for any damage that may have been caused.

2.1.2 Extra-contractual liability

This liability is arises from the commission of a civil offence or quasi-offence, i.e. an offence or quasi-offence which gives rise to the obligation by the person who has committed it to compensate for the damage thus caused to the affected party or parties.

An *offence* is an unlawful, fraudulent, intentional act that causes damage to another, while a *quasi-offence* is an unlawful negligent act that causes damage to another. The difference between offence and quasi-offence therefore relates to intention in the case of an offence and culpability in the case of a quasi-offence.

2.1.3 Legal Liability

Legal liability is liability emanating directly from the law (For example (CC Art. 222): 'Parents, in concert, or a surviving father or mother, are responsible for personal care in raising and educating their legitimate children').

2.1.4 Criminal Liability

Like extra-contractual liability, criminal liability originates in the commission of an offence or quasi-offence, but with the difference that it must be criminal.

A criminal offence is any voluntary action or omission punishable by law, while a criminal quasi-offence is any culpable action or omission punishable by law.

3 The vendor's contractual liability

The execution of a purchase-and-sale agreement creates rights and obligations, both for the purchaser and for the vendor.

The principal obligation of any purchaser is payment of the price, while those of the vendor are delivery of the thing sold and giving good title to the same.

The giving of a good title to the thing sold includes two objectives: clearing of title and clearing of redhibitory (below) defects. The clearing of title refers to the obligation of the vendor to protect the purchaser in terms of ownership and peaceful enjoyment of the thing sold. The clearing of redhibitory defects refers to the vendor's obligation to answer for hidden defects in the thing sold.

The situation raises the issue of liability for redhibitory defects, i.e. the vendor's obligation to procure the purchaser's useful possession of the thing sold, in a condition useable for the purposes that led to the acquisition.

When there are hidden defects in the thing contemplated in the purchase-and-sale, CC Art. 1857 entitles the purchaser to exercise one of the following judicial actions against the vendor:

- Redhibitory action, whereby he may judicially request rescission or voidance of the purchase-and-sale;
- *Quanti minoris* action whereby he may judicially request that the price be lowered.

If two or more things are purchased and only one has hidden defects, the purchaser may not file such actions except regarding the thing sold with the defect, the sale continuing regarding the other goods, save the case where the purchaser would not have purchased the whole without the thing with the defect.

The legal requirements for an action for redhibitory defects or a *quanti minoris* action are:

(i) the defect must be hidden;

(ii) the defect must have existed at the time of sale;

(iii) the defect must be so serious that it may be presumed that, if known to the purchaser, he would not have purchased or would have done so at a much lower price;

(iv) the judicial action must not have been extinguished.

Additionally, a vendor who has acted in bad faith is liable for loss caused to the purchaser. The vendor is taken to have acted in bad faith if it knew of the defects in the thing, or if it should have known of them because of its profession or trade, and did not declare them. A claim for damages for bad faith by the vendor is not an action parallel to redhibitory and *quanti minoris* actions, but is rather part of it, so that if such actions are lost through prescription, the claim for damages will also be subject to pre-scription.

We will explore only matters of interest to a solution of the case in question.

(i) The defects must be hidden, meaning that they were not indicated by the vendor and were in such a form that the purchaser was able to ignore them without serious negligence on its part or in such a form that the pur-chaser could not have easily known thereof because of its profession or trade.

Hence, defects will not be considered hidden if the vendor discloses them to the purchaser, if the non-expert purchaser has ignored them due to serious negligence or if the purchaser could have easily recognised them due to its profession or trade.

Case law has largely determined that there is no serious negligence if the purchaser has not had the thing purchased inspected by an expert or technician, since the purchaser is only required to make a personal inspection, which, according to a segment of case law, does not need to be exhaustive; this is because negligence on the part of the purchaser must be serious. Superior Courts have declared that if the thing purchased is apparently defect-free, serious negligence cannot be attributed to the purchaser for omitting an expert survey of the thing in a purchase implemented under normal conditions in the particular type of business.[3]

In any case, serious negligence on the part of the purchaser will be determined by the judge.

(ii) The defects must have existed at the time of sale.

The vendor will be liable only for defects existing at the time of the contract and for those supervening at a later date.

Both case law[4] and doctrine[5] have largely declared that there is no requirement that the defect existed in all its gravity and magnitude at the time of sale: it suffices that only the 'seed' or potential exists, meaning that even when the effects of a defect have not become patent, they originate in a circumstance or characteristic that was present at the date of the contract.

(iii) The defect must be serious.

The requirement relative to the gravity of the defect, depending on the same, shall determine whether an action to rescind the purchase and sale – a redhibitory action – or an action to lower the price – a *quanti minoris* action – is warranted.

If the defects are so grave as to make the thing useless for its natural purpose or imperfectly usable so that it can be presumed that the purchaser would not have purchased it, a redhibitory action will be warranted, and if they are so grave as to cause the presumption that the purchaser would have purchased at a much lower price, a *quanti minoris* action will be warranted. The courts are restrictive in granting rescission, since they give priority to the finality of transactions:

> If a collapse of the walls in a house purchased does not make it unsuitable for its natural use although repairs are required, only the price will be reduced, but the contract not rescinded.[6]

In order for an action to lower the price to be justified, not only must it be presumable that the purchaser would have purchased at a lower price, but also presumable that it would have purchased at a much lower price. The criteria used to determine what the lower price would be to warrant a *quanti minoris* action are a matter of fact to be determined by the court.

(iv) A judicial action must not be extinguished.

One last requirement for an action for redhibitory defects is that such action has not been extinguished. Since an action may be extinguished on several grounds, none of them must have taken effect.

It must not be lost through prescription: Chilean law establishes terms for prescription for the filing of both a redhibitory and a *quanti minoris* action.

These terms regarding real estate are one year from the delivery of the thing for a redhibitory action and 18 months for a *quanti minoris* action.

Statute did not establish the point from when the periods begin to run, but case law has declared that it is considered from the delivery of the good to the purchaser:

> The prescription of an action to lower the price (a *quanti minoris* action) as well as an action to rescind the contract begins to run as of the material delivery, the date from which the purchaser, having the thing in his power, may easily verify the defects thereof.[7]

However, this raises an interesting question, as special rules are applicable with priority over general rules. Many matters regulated in the Civil Code are also regulated more specifically by the Commercial Code, so if we find rules in the latter relative to construction, they must take priority over those we have previously mentioned.

In regard to the issue of redhibitory defects in purchase-and-sale contracts, the Commercial Code Art. 154 states that

> the vendor is obligated to remedy the merchandise sold and to be liable for hidden defects contained therein pursuant to the rules established in the Title on Purchase and Sale in the Civil Code.
>
> Redhibitory actions will be lost through prescription after six months from the day of real delivery of the thing.

This rule is applicable when the transaction, in this case the sale of the building, is commercial in nature. In fact, the Criminal Code Art. 1 establishes that its rules govern the obligations of merchants referring to mercantile transactions which are assumed by non-business people to secure compliance with commercial obligations, as well as those resulting from exclusively mercantile contracts. And the Commercial Code Art. 2 adds: 'Whenever cases are not specially resolved by this Code, the provisions in the Civil Code shall be applicable.'

The Commercial Code Art. 3 establishes a list of commercial acts:

Commercial acts, either by both contracting parties, or by one thereof. ...
Construction companies of immovable... such as buildings, roads, bridges, canals, drains, industrial facilities and other facilities of the same nature.

By virtue of this rule, construction activity is defined as commercial, and the rules of the Commercial Code are applicable:

Consequently, in order for the economic activity of construction of immovable property to be included in acts that the law considers mercantile, such activity must be exercised under the organisation of a company, meaning by means of a set of material and human means gathered together by one person and co-ordinated for the purposes of participating in the market for goods and services.[8]

What happens with the sale of a building that was built by a construction company in the course of its business? Although it is not commercial in origin, since only construction activity in and of itself is, it must also be given this nature by virtue of the *theory of adjunction* which

consists of presuming certain acts to be mercantile when they relate to a profession, activity or principal legal act commercial in nature, either because they facilitate [or] contribute to its growth or execution or simply guarantee.[9]

This criterion is fully accepted in doctrine and applied by the courts.

The following scenarios must be taken into consideration for application of this theory: the developer (vendor) commissions the construction of a work by means of a lump-sum contract, and the contractor is the one to supply all or most of the materials; the work is its work, and upon completion thereof it sells the same to whoever commissioned the work. If the contractor is the one to execute the commercial act of construction, and the sale that is made of the work to the real estate company is mercantile in nature because it is an act accessory to the principal activity, that is mercantile in and of itself. Later, the sale of the work by the developer is civil, since the theory of adjunction only arises from the sale by the contractor.

If, however, the developer had commissioned the construction of the work by means of a 'management contract' – i.e. it supplies all or most of the materials and is the owner of the work from the beginning but contracts for the services of the contractor –, then the developer is performing a commercial act from the beginning, and the sale of the work is mercantile under the theory of adjunction, since it contributes to achieving the purpose of the principal activity (profit) that is part of this, and which is commercial. And if one and the same corporation fulfils the roles of developer and construction company, the sale of the work is also commercial for the same reasons.

The qualification of the relationship between the client and the contractor arises under CC Art. 1996:

> If the craftsman supplies the materials for manufacture of a material work, the contract is a sales contract; however, it is not perfected except by approval of whoever ordered the work.

And Sub-paragraph 3:

> If the material is supplied by the person who commissioned the work, the contract is a lease.

In these scenarios, the sale of the work by the developer to the investment bank is a mixed act: mercantile in nature for the vendor and civil in nature for the purchaser. The theory of adjunction applies only regarding the act of the vendor, since for him the sale is preceded by an act that is clearly commercial – the construction of the property –, which does not occur with the purchaser. Both doctrine and case law consider that the law of the obligor is applicable in mixed acts: 'Legislation corresponding to the person obligated according to the status of the act therefore shall be applicable.'[10, 11]

In this case, since the person required to give good title is the vendor – for whom the act is commercial in nature –, the Commercial Code Art. 154 would be applicable: the term of prescription of actions is six months from the delivery of the thing, and not the term of one year or 18 months in the Civil Code.

Is it possible to apply the theory described whereby the term of prescription of redhibitory actions is six months as established in Commercial Code Art. 154? A doubt arises because sub-paragraph one of this rule speaks of 'merchandise,' and traditionally this word has been interpreted to mean chattels, so the rule is applicable only to the purchase and sale of chattels. Moreover, that is the meaning that is given to it in the dictionary of the Royal Academy of the Spanish Language.

In order to defend the application of this theory, the vendor – for whom the shorter term of prescription as established in the Commercial Code is convenient – therefore may argue that when such legal wording was issued, the word 'merchandise' was understood as 'chattels' since at that time only the transactions referring to chattels were considered to be commercial. However, when later (in 1977) the activity of construction was included among commercial acts, a change in the meaning of the word 'merchandise' would have taken effect in all rules of the Commercial Code, including Article 154, making it applicable to immovable property.

In our opinion, the court would apply the terms established in the Civil Code, provided the rules of the Commercial Code are applicable in general only to transactions involving chattels, giving a restrictive meaning to the expression 'merchandise.' We have included the theory described because it is possible that it may gain ground in the future, since the intent is currently observed in doctrine to expand the application of commercial rules to transactions referring to real estate.

The prescription may have been waived by the vendor, but this would have an effect only if the vendor acted in good faith (above).

The sale must not be involuntary: According to CC Art. 1865, redhibitory and *quanti minoris* actions are inapplicable to compulsory acquisitions made by the public authorities or by order of a court, i.e. sales where the purchaser has made the purchase in a judicial action.

In accordance with CC Art. 1863, the parties may, in contract, assign 'redhibitory' to certain defects that by nature are not redhibitory, i.e. they may establish that certain defects that do not meet the legal requirements listed above to be redhibitory will be redhibitory for any purpose.

4 The vendor's extra-contractual liability

As mentioned, extra-contractual liability originates in the commission of a civil offence or quasi-offence.

In the case considered, the vendor has committed an act (the sale in which the thing sold had a hidden defect) that has caused loss of benefit to the purchaser.

In view of the foregoing, even though the vendor is tied to the purchaser by a contract from which contractual liabilities emanate, the same negligence in contract performance that gave rise to such negligence may constitute a civil offence or quasi-offence, generating extra-contractual liabilities, and therefore the obligation to pay compensation for damage caused.

As a result, liabilities are duplicated, namely the contractual liability under the contract and extra-contractual liability under the intentional or culpable default on obligations arising under the contract from which damage originates to the other party to the contract.

(a) Accumulation of liabilities

This duplication of liabilities has been called an 'accumulation of liabilities' by doctrine. The issue of an accumulation of liabilities consists simply of determining whether the violation of a contractual obligation may give rise to contractual or extra-contractual liability without distinction or only to contractual liability, that is, whether the damage arising from such violation entitles the creditor to choose between both liabilities and sue for compensation according to whichever it finds most convenient.

The accumulation of liabilities has been rejected by doctrine and by case law, both declaring that in the presence of a contract, extra-contractual liabilities are impossible due to default on the contract, and only the rules on contractual liability are applicable.

The Supreme Court has therefore resolved that

> Article 2329 of the Civil Code, which prescribes as a general rule that any damage that may attributed to maliciousness of negligence of another person must be redressed thereby, is applicable to the case where the obligation to compensate arises from a criminal violation and not a violation of any contractual obligation, such as where the plaintiff accuses the defendant of not having exercised an action within the appropriate period of time.[12]

Notwithstanding the foregoing, certain rulings by our Superior Courts have accepted an accumulation of liabilities.[13]

(b) Extra-contractual liability for acts of employees

CC Art. 2320 provides that any person is liable not only for its own actions but also for the acts of those who were under its care, and CC Art. 2322 establishes that masters shall be liable for the behaviour of their servants or domestic personnel in the exercise of their respective functions; the concept of domestic personnel or servant has been extended by case law to employees or workers in a productive company.[14] This liability does not affect the vendor as such, but rather emanates from an unlawful act by its employees and from the tie of dependence between it and the persons that participated in the construction.

In order to attribute liability to the vendor for the acts of its employees, the unlawfulness or illegality of the acts of the persons who participate in the construction and the tie of employment binding the vendor thereto must be determined.

Therefore, in order for such liability on the part of the vendor to be established, proof of two elements is required: the first relates to the commission of offences or quasi-offences by a person who participated in the construction, and the second to the relationship of this person with the vendor.

'Employees' has been defined by doctrine as 'persons who are at the service (of another) such as employees, workers, etc. What characterises the employee is the fact of being subordinated to another person, working under the authority or orders of another.'[15]

Case law has decided that in the case of a lump-sum contract between whoever commissions the work and the parties that participated therein, there is no tie of dependency as such. In a lump-sum construction contract, the contractor contributes all construction materials which, upon completion, are sold to the client for a single total amount. The Supreme Court therefore resolved that

> the fact that the civil plaintiff contracted work from the defendant that should have been performed on property owned thereby for a lump-sum price, and the fact that it provided the producer [*constructor*] with materials for construction of the work do not necessarily lead to the establishment of a dependency by the latter on the former. Dependency assumes subordination, and in this matter it cannot be seen how a contractor executing work for a lump-sum price is subordinated to whoever ordered the manufacture of the same.[16]

On the other hand, if the construction contract binding the vendor to the subjects participating in the work is a management construction contract (i.e. whoever commissioned the work contributes all materials and only contracts the services, cf. above), a legal framework would exist where it is possible for the legal requirements to be fulfilled, creating a relationship of dependency between whoever commissions the work and the subjects participating therein.

The position may be supported by doctrine, which has upheld the view that

> the status of employee is not incompatible with the fact that the job or position performed requires technical knowledge unknown to the principal: professionals – attorneys, doctors, engineers, etc. – may be employees of a businessman ... If they work under the orders thereof ... and if they commit an offence or quasi-offence while they are under the care of the businessman, the latter is liable.[17]

(c) Requirements for and elements of extra-contractual liability

The requirements or elements in extra-contractual liability for a claim for damages to be justified for the commission of a civil offence or quasi-offence are the following:

(i) that the affected party has suffered damage – the loss of a right or deprivation of a legitimate advantage;

(ii) that the party in breach has acted intentionally – intending to cause damage – or negligently;

(iii) that there is a cause-effect relationship between the wilful misconduct or negligence and the damage, meaning that the damage is the result of the wilful misconduct or negligence of the offender;

(iv) that the offender has a criminal capacity, meaning that it is a person capable of discerning between the meaning and consequence of its acts;

(v) that the judicial action is prescribed (the term for prescription to file a claim for damages for the commission of an unlawful civil act is four years as of the date of the event);

The fundamental differences between contractual and extra-contractual liability from the viewpoint of the plaintiff's rights are as follows:

(i) Terms of Prescription: The terms of prescription in the case of contractual liability for hidden defects in the sale of real estate are, as mentioned, shorter than the terms of prescription for extra-contractual liability;

(ii) *in solidum* liability: If there are several persons extra-contractually liable for the unlawful events, they are jointly and severally liable, i.e. the affected party may sue any of those liable for all of the damages. In contractual liability, liability is simply joint, meaning that each person is liable for the proportion of the damages corresponding thereto, depending on the number of defendants. This rule may be changed in the individual contract, in which joint and several liability is expressly agreed upon.

5 Producers' extra-contractual liability

Our legislation establishes special rules regarding the liability of producers (professionals and trades) involved in the construction. The *General Law on Urbanism and Construction* (GLU&C) therefore expressly, in its Article 18, provides that

> Manufacturers, designers and contractors are liable, respectively, for the quality of the materials, errors in design and defects in construction in the works in which they have been involved and the damage that is caused to third parties thereby.

Most doctrine says that this rule establishes an objective extra-contractual liability for producers, so whoever suffers damage does not have to prove all elements of extra-contractual liability, but rather only the damage caused, in addition to the involvement of the producers, and that the damage is the result of their acts: there is no need to prove negligence or wilful misconduct.

It is important to add that in accordance with the stipulations of the GLU&C Art. 124, 2nd paragraph, bodies corporate organised as construction companies or as designing companies shall be jointly and severally liable, together with the competent producer acting on behalf of them, regarding the defects in construction or errors in design, respectively, in the works entrusted to them and the damage caused to a third party.

6 Extra-contractual liability of sub-contractors

Like the vendor, sub-contractors are liable for the civil offences or quasi-offences they have committed in the construction, and if such a sub-contractor is a company, the company shall be liable for the acts of its employees, provided the requirements mentioned are present (the commission of an unlawful act by the employee and a relationship of dependency between them and the sub-contractor).

If the sub-contractor is a manufacturer, architect or contractor, the rule on professional liability contained in GLU&C Art. 18 is applicable.

7 Liability of suppliers of materials

In the case under study, the vendor of the material that caused the damage resides in a foreign country, so it is impossible for the purchaser to establish its liability. However, we will discuss this point according to national rules.

GLU&C Art. 18 states that the manufacturer shall be liable for the quality of the materials causing defects in the works in which they were used if loss is caused to a third party. This is the same liability as the one to which producers involved in the work are subject, so there is no need to demonstrate that the manufacturer acted intentionally or negligently: even if he did not know of the defect in the material, he must bear responsibility for damage arising from it. The damaged party has a period of five years from acceptance of the work by the municipality in which to file a suit.

8 Municipalities' extra-contractual liability

Although the case under study does not contemplate this possibility, extra-contractual liability of the Municipality may arise under Chilean law.

Municipalities are organs of the State Administration in charge of governing the townships into which the Chilean territory is divided. Their attributes include inspecting construction works within the township. GLU&C Art. 142 expands upon and specifies this attribute:

> Once a work or part thereof that may be fitted out independently has been completed, the owner or supervisor shall request final inspection thereof and final acceptance thereof by the Municipal Works Office.

Only once final acceptance has been issued by the Municipality may the work be fitted out or used (Article 145):

> No work may be inhabited or used in any way before its partial and total final acceptance.

Municipal acceptance of the construction work guarantees the producers, the purchaser and any person that such work is in a condition to be inhabited and used. Therefore, if the Municipality accords final acceptance without the work being suitable for its use – negligently –, the damages arising, for whoever purchases or uses it, are imputable to this body. Moreover, the Constitutional Charter of Municipalities Art. 137 explains it further:

> Municipalities shall be liable for the damage they cause which shall arise mainly due to a lack of service.

Although in the case under study we are not concerned with a lack of service, this is not the only ground for the liability of Municipalities, but rather only the principal ground. Nonetheless, this does not exclude liability derived, for example, from negligent service or complying negligently with a duty imposed by law.

In conclusion, if a Municipality accepts works that are later proven not to meet the conditions for their natural use, it incurs extra-contractual liability according to the general rules of the Civil Code. In Type B-I (below) we describe a case where the plaintiff sued and obtained judgment against a Municipality that had granted a final approval of a building which did not resist an earthquake.

9 Criminal liability of vendors, producers and sub-contractors

In the case under study, it is possible to configure the criminal liability of the subjects involved in the construction, both regarding the vendor and the producers. In a concrete case, the criminal act that may be contemplated is the criminal act generically called 'fraud by deceit.'

The elements comprising the criminal act of fraud by deceit are as follows:

(a) *Simulation*: Is activity undertaken by the subject which may consist of any action or omission that may create in another a misrepresentation of a reality? A simple lie does not suffice; external appearances or certain exceptional circumstances that accompany the mendacious affirmation must exist;

(b) *Error*, i.e. a misrepresentation of reality by the other party;

(c) *Equity disposition* exists if the intentional act causes a diminution in the equity thereof;

(d) *Damage*, i.e. damage or impairment to the equity of a person;

(e) *Cause-effect relationship*, i.e. a causative tie must exist between all elements indicated above.

10 Circumstances or elements that release producers from liability

The circumstances that will release the subjects involved in construction from liability consist fundamentally of decreasing the presence of the requirements constituting each of the types of liabilities in the preceding sections.

10.1 Regarding the vendor
10.1.1 Contractual ambit
The circumstances that will release the vendor from contractual liability consist mainly of the following:

As indicated, the first element that must be present in order to generate contractual liability due to redhibitory defects is that the defects of the thing sold must be hidden. This presupposes two requirements: That the vendor has not declared them, and that they are of such a magnitude that the purchaser has been able to ignore them without gross negligence on its part, or they are such that the purchaser has not been able to recognise them easily by reason of his profession or trade. The fact that gross negligence did or did not exist on the part of the purchaser is a matter of fact which the court must appreciate, so the vendor must prove that the purchaser committed gross negligence by not inspecting the goods with the diligence normally to be expected, which will vary with the type of goods in question.

The second element needed to establish contractual liability for redhibitory defects is that the defect must exist at the time of sale. As mentioned, it is not necessary for the defect to exist in its fullest form; a 'seed' or the potential for damage will be sufficient. The vendor must prove that the defect did not exist in any of its development stages, and that even though doctrine and case law state that it suffices for the 'seed' to exist, this presupposes that it should have been manifest in at least some form, and that the event or condition that caused the thing to make manifest its defects are subsequent to the sale.

The third element relates to the gravity of the defect in regard to the fact that if the purchaser had known of it, it would not have bought the same (in the case of a redhibitory action), or it would have purchased it at a much lower price (in the case of *quanti minoris* action). The vendor must prove that the defect affecting the thing is absolutely accidental and does not have the gravity necessary, as established by law, in order for the sale to be rescinded or the price to be lowered.

Lastly, it must be shown that the time to file the redhibitory or *quanti minoris* action has not elapsed. The term for prescription of both actions runs from the date of delivery of the thing sold, regardless of when the defect became evident. It may therefore happen that the defect becomes evident only once the right to action has been lost by prescription.

10.2 Extra-contractual ambit

First of all, in view of an action by the purchaser to attribute extra-contractual liability, the vendor would diminish its liability and adduce the existence of a contractual tie, and therefore a lack of justification of extra-contractual liability based on the doctrine and case law that rejects what has been called accumulation of liabilities above.

In addition to the defence referred to above, the defendant must diminish the presence of the elements or requirements for extra-contractual liability. Principally, the presence of wilful misconduct or negligence should be negated.

Now, if the liability to be attributed to the vendor is not for an act thereof but rather for an act of an employee thereof involved in the work, the evidence should also focus on proving the absence of the requirements linking the liability of such subjects to the vendor.

Additionally, and in the case under study, the vendor would evidence that it did not have the means to prevent or impede the unlawful actions by the persons who participated in the construction, all in accordance with the stipulations of the CC Art. 2322 (final sub-paragraph), which provides that masters shall not be liable for what their servants or domestic personnel have done in the exercise of their respective functions if it is proven that they have exercised them improperly, that the masters did not have the means to prevent them using less than regular care, nor the competent authority, in which case any liability shall fall upon the domestic personnel or servants. As indicated, case law has expanded the concept of domestic personnel or servant to any dependent person in the exercise of their functions.

10.3 Penal ambit

Any criminal liability of the vendor shall depend upon the joint and successive presence of all elements for the criminal act of fraud by deceit. Therefore, the defence must be aimed at diminishing the presence of one or more of such requirements.

So it must be evidenced that no simulation existed, i.e. that the silence of the vendor in the making of the contract with the vendor regarding the defects of the thing contemplated in the contract was not an intentional action on its part (that there was no intent for the purchaser to form a misrepresentation of the features of the thing).

11 Producers' liability

11.1 Extra-contractual ambit

As indicated above, extra-contractual liability of professionals is determined by law (GLU&C Art. 18). In order to be released from liability, the contractor and the architect must prove that their involvement and actions in the construction were performed as diligently as should a producer, and that the defects in the work did not arise from the acts for which they are liable, namely, defects in design and in construction.

11.2 Penal ambit

As for the vendor, the criminal liability of producers depends on the joint and several presence of the requirements and elements of the criminal act of fraud by deceit, and such element should therefore be diminished.

Type A

THE PURCHASER'S OPPORTUNITY FOR RECOVERY, THROUGH LITIGATION OR OTHERWISE, FROM THE VENDOR

The investment bank (the Bank) has a series of alternatives to recover the loss suffered from the vendor. These alternatives are based on different legal ambits, each of which will be explained separately.

1 Contractual aspects

The legal nature of the relationship between the vendor and purchaser is the relationship of a purchase-and-sale contract that generates rights and obligations for the parties.

The rights available to the purchaser include, *inter alia*, the right of 'delivery of the thing sold' and the right that it 'be useful for its natural purpose' (CC Arts 1824 ff.). The principal right of the purchaser that may have been violated is that the thing sold is not useful for its natural purpose.

In fact, as described above, the purchaser is entitled to demand that the vendor rescind the sale or proportionally lower the price due to hidden defects in the thing sold, whether immovable or chattels, called redhibitory defects. Such rights, which give rise to the so-called 'redhibitory action,' are established in CC Art. 1857, cf. *Introduction*. In this regard, our principal concern is to resolve whether they are applicable to the case presented.

First, That the defect existed at the time of sale
As the case is stated, the defect is hairline cracks in the tiles becoming patent after the property was acquired but obviously not present at the time of purchase.

At the time of purchase, the tiles did not show the cracks that in the end prevented the Bank from renting out the units, but the risk of their deterioration was latent since their composition was such that they tended to develop the cracks, as evidenced by the technical report. It is inferred from the foregoing, with a certain clarity, that the 'seed' of the defect did exist at the time of the purchase-and-sale contract, although not perceptible in all of its magnitude and gravity.

We noted above that this point has been clearly resolved largely by doctrine and by case law, in the sense that it suffices for the seed of defect to exist at the time of sale in order to give rise to a redhibitory action.

The first requirement that the defect existed at the time of sale is fulfilled for an action by the Bank against the vendor.

Second, That, because of the defect, the thing is not or less than fully fit for its natural purpose, so that the purchaser, had it known of such defect, would not have made the purchase or would have done so at a much lower price, meaning that the defect is serious.

The Bank acquired the property for the exclusive purpose of leasing the units, in other words a real estate investment specific to the Bank's business.

As it will be impossible for the Bank to fulfil its objective – or it will be able do so only imperfectly –, it is reasonable to think that the Bank could sue for rescission of the contract. However, CC Art. 1868 provides that

> If hidden defects are not of the degree of significance expressed in paragraph 2 of Article 1858, the purchaser shall not be entitled to rescind the sale but rather merely to lower the price.

Consequently, it needs to be determined whether the defects are sufficient to provide an action for rescission, or, if not, for the price merely to be lowered.

The defect does not affect the natural use of the property, since the defect is solely decorative, although the Bank would use it for a purpose that, given the failures, will notably affect its performance. So the logical thing is to decide what the meaning of 'natural use' is, so that in a future redhibitory action, the Bank may argue that the natural use of the property corresponds to the use that the Bank intended to give it.

The criterion used by the courts regarding this point is single-minded, since claims and litigation based on this type of situation create instability in transactions. Therefore, in order for the rescission to be successful, one must be able to demonstrate the impossibility of the natural use of the thing and the gravity of the damage.

In our opinion, the natural use of property arises from an objective condition more than from the subjective use intended by the purchaser, notwithstanding that an aesthetic defect must be resolved by means of an agreement, at a price. We are therefore inclined to think that the Bank could sue and obtain a reduction in the agreed price in terms of the replacement cost of the tiles.

Third, That the vendor had not indicated the defects and that the purchaser was able to ignore them without committing gross negligence or was unable to recognise them easily by reason of his profession or trade, meaning that the defect is hidden.

In order to fulfil this requirement, the purchaser will have to demonstrate that it could not have known of the defects, that they were not reported when appropriate, and that its profession or trade did not permit it to know about them. In our case, the Bank will need to prove such conditions; however, in the light of the account given for this case, it will be very simple for it to prove its ignorance of the defects by reason of the lack of information provided by the vendor, that this ignorance was not negligent and that its trade did not give it the tools necessary to have known of them. As mentioned, case law has resolved that gross negligence cannot be attributed to the purchaser for failure to have an expert appraisal of the thing when the purchase has been made as is customary for the type of business. Consequently, the Bank is in a very advantageous position, since it can demonstrate, quite easily, that the defect was hidden.

However, there are additional elements here that may compound the vendor's difficulties. In fact, as provided in CC Art. 1861, if the vendor acted in bad faith, i.e. if it knew of the defects and did not declare them, or if they were of such a nature that it should have known of them by reason of its profession or trade, it will also be obliged to pay compensation.

In the case under study, if the vendor knew of the technical report at the time of the sale, it would also be required to pay compensation. Liability may be placed upon the vendor if it ought to have known of the report by reason of its profession or trade.

If the purchaser can demonstrate that the vendor was aware of the technical report, it would also have the right to demand damage compensation, in addition to the *quanti minoris* action.

Additionally, if it could be demonstrated by the Bank that the architect in charge objected to the use of this material, this would reinforce the arguments that the vendor acted negligently by not adopting measures to ensure the suitability of the material, and that it acted in bad faith because the architect advised it of the risk of using the tiles.

Therefore, the Bank has a clear prospect of securing damage compensation as provided by CC Art. 1861. It thus appears that the Bank can recover the losses it has suffered by these means.

Fourth, One has to consider whether judicial action has been extinguished.

As indicated in the *Introduction*, in our opinion, the possibility of applying the term of prescription of the Commercial Code Art. 154 (six months after delivery of the thing) must be discarded because, according to the literal text of this rule, it applies to the purchase and sale of chattels.

Accordingly, the general rules of the Civil Code are applicable. As mentioned, these provide a term of one year from the material delivery of the product to request rescission of the contract regarding real estate, or redhibitory action (CC Art. 1866); and they also contemplate a period of 18 months to request a reduction in the purchase price of the property (CC Art. 1869); for a *quanti minoris* action, unlike the redhibitory action, statute does not say when the term begins to run, although case law has decided that the term begins at the time of the material delivery of the thing.[18]

Consequently, the Bank will have the period of one year to file a suit against the vendor for rescission, and of 18 months to lower the price. According to the facts provided, the Bank has perceived the defects quite rapidly, so it is possible to consider that the time to file either or both of the two actions (but one subject to the other), is currently available.

Furthermore, as noted above, the Bank has not waived its right of action, nor is it a question of an involuntary sale, so the formal requirements to sue are fulfilled.

2 Extra-contractual aspects

Notwithstanding the foregoing, and in the event that the Bank does not have an action against the vendor because such actions have been lost through prescription, the Bank may file actions against the vendor via extra-contractual liability. It is at this point important to clarify that extra-contractual liability is lost by prescription in a period of four years, so the Bank may bring an action although the periods specific to actions for hidden defects have expired.

As noted above, case law recognises that the theory sustaining a contract creditor's right to claim compensation for damage suffered by default on the contract because the contract debtor has committed a civil offence or quasi-offence, arises pursuant to the rules governing extra-contractual liability. In this regard, we noted the way in which this so-called accumulation of liabilities must be confronted in order for it to be sustained by the court. As a result, for this particular case, we will refer only to the concrete measures to be adopted by the Bank to make a claim against the vendor.

In the case under study, we shall endeavour to demonstrate all elements essential to extra-contractual liability. In summary, we need to prove the existence of actual damage, that it was caused by whoever appears as our possible defendant, and lastly that the actions of the defendant were such that the damage indicated was caused by its gross negligence or wilful misconduct.

First of all, the Bank can demonstrate quite easily the existence of damage. In fact, by purchasing the product for investment and then trying to rent it out, the Bank suffered a clear loss because of the building's decayed appearance.

Although the vendor will be able to argue that such damage is not real, since the property can be used for its normal purpose, the mere demonstration of the hairline cracks in the tiles will prove the existence of damage.

The Bank must then prove that the damage was caused by the vendor's actions. At this point, the Bank should prove that the vendor was exclusively liable for the production of the building, and that the persons involved acted on its behalf. To do so, it will be important to know whether the contractor in charge of the execution of the work and the designing architect were employed by the vendor or acted independently.

Notwithstanding the foregoing, it will be decisive for the Bank to be able to prove that the decision to use this type of tiles was exclusively the vendor's, and that, although there is a tie to the contractor or architect leading to the presumption that they directed the construction independently, the mere fact of its having imposed the choice of tiles makes it the cause of the damage.

Finally, the Bank must prove that the vendor's conduct was negligent or intentional, and that the vendor did not conduct itself prudently as would a counterpart in the same profession or trade, located in the same place, at the same time and under the same external circumstances. It must prove how a judicious person would have acted, what concrete measures such a person would have adopted, and under what conditions.

To demonstrate that the actions of the vendor create liability, it would be desirable to prove, by the architect's testimony, that the developer received a warning about the use of an unknown material, a fact that will obviously show more precisely its degree of liability. As things are, if the Bank can demonstrate that the purchase of the tiles was also due to a decision based on the cost of the work or another element – such as, for example, economic interests in the manufacture of the tiles – that are in opposition to prudent and professional conduct, it will be in a position, and have the opportunity, to prove by the events that the developer acted negligently.

If it were possible to gather together the above elements in a precise manner, it would be possible to bring an action for the extra-contractual liability of the vendor, even though there is a purchase-and-sale contract in the midst, but always on the understanding that this type of action will be grounded in a theory that has supporters and detractors in our courts.

As a corollary to the above, a brief comment must be made on the liability of the vendor for the acts of its employees.

In this case, the following distinctions will have to be made:

- Whether the decision to install the tiles was adopted directly by the developer;
- Whether the installation of the tiles was adopted by the architect or by the contractor or by both.

As to the first point, if this was the developer's decision, then it will not be material to review the legal tie joining it to the producers.

As to the second point, if the answer is affirmative, then it will be decisive, concerning the possible extra-contractual liability, to know what the legal tie was between them. In effect, assuming that the decision was adopted by one of the producers, it will have to be resolved whether or not they acted independently of the vendor. In such an event, the type of contract binding the parties will need to be studied.

It therefore follows that if there is a lump-sum contract between the developer and the architect or the contractor, the producers are acknowledged to be fully independent, so the decisions adopted in performance of their assignment makes only them liable for the same; there would be no *in solidum* liability which would make the developer liable. Furthermore, if the connection is through employment for labour or management (see *Introduction*), or by any other means signifying that the control of the work remained with the developer, the developer will be jointly and severally liable for the acts of its employees, and in such case, the suit for extra-contractual liability against them will be justified, as described above concerning extra-contractual liability for acts of employees.

3 The convenience of filing one action or the other

It will be advantageous for the Bank to bring the action relative to hidden defects, because this does not affect the terms for prescription. There is no need, in this case, to demonstrate any guilty act by the vendor, but merely the existence of the contract, the damage and the liability pertaining to it. The Bank need not prove negligent or intentional conduct, except in order to show aggravation of liability.

Moreover, the action arising from extra-contractual liability will be useful to the extent that the terms for redhibitory or *quanti minoris* action have been lost by prescription. It will be advisable only if this has happened.

4 Penal ambit

Whereas the penal means were outlined in the *Introduction*, it should be decided whether a criminal complaint can be filed against those proving to be responsible for the offences indicated and consequently secure compensation for the damage caused, which is the goal sought by the Bank.

According to the foregoing, the Bank would file suit for fraud by deceit, which will obligate it to demonstrate that the vendor, knowing of the research on the tiles, kept silent and so misled the Bank. The vendor's simulation of the real events was the cause of error by the Bank.

This type of action is advisable when there is a good likelihood of obtaining a favourable ruling, since, in such a case, the defendant prefers to begin negotiations before being condemned; also, in view of other problems, given the circumstances, the Bank would be in an advantageous position. However, one must consider that this offence intrinsically bears a relatively minor degree of punishment by imprisonment, although the importance lies in obtaining recognition of the damage caused to the plaintiff and the declaration of the developer as civilly liable by a favourable ruling, which makes this alternative seem to be quite recommendable if the assumptions and evidence are handled correctly.

THE PURCHASER'S OPPORTUNITY FOR RECOVERY, THROUGH LITIGATION OR OTHERWISE, FROM THE VENDOR'S CONTRACTOR

An analysis should first be made as to whether a claim can be made against the contractor, and compensation obtained.

There are two possible ways to respond to this question:

(a) Suing the contractor for its professional liability
As stated in the *Introduction*, Art. 18 of the General Law on Urbanism and Construction (GLU&C) establishes objective liability, i.e. liability is assumed for the damage and the involvement in causing it, once this has been proven, notwithstanding the fact that the conduct has been neither negligent nor intentional.

A study should be made as to whether the contractor did or did not fulfil its professional liability as required by the norm. What would its professional liability be? According to custom, the contractor is liable for inspecting the materials to be used, getting right the proportions and the mix of materials, demanding evidence, necessary analyses and suitably precise documents for obtaining approval.

Proof will be required of the damage and the contractor's involvement in causing this for an action against it to succeed; however, if it can be proved that the contractor's actions were prudent and it adopted a professional attitude, the liability attributed to it will be diminished and the suit against it will fail.

(b) Suing the contractor for its extra-contractual liability pursuant to the rules of the Civil Code.
In relation to the contractor, the Bank's alternative will be to institute actions against the contractor for extra-contractual liability, in which the Bank must prove not only the contractor's involvement but also its wilful misconduct or negligence. The only advantage to bringing an action of this nature is that it contemplates the joint and several liability of those involved, even when it must be proven, from the standpoint of evidence, that the participants acted intentionally or negligently.

Concerning the terms allowed to bring each of the above actions, it is important to bear in mind that the terms of the GLU&C are five years as from final approval of the work by the Municipality, and the terms for extra-contractual liability are four years from perpetration of the event, so the terms for professional liability are clearly advantageous. Accordingly, save for solidarity, the action against the contractor should be under the GLU&C.

In turn, if a tie between the contractor and the developer can be demonstrated, then it would seem more advisable to sue the contractor and the developer jointly and severally, as established in the preceding response. This means suing the contractor jointly and severally as a participant in the event by having a tie to the contractor that makes it jointly and severally liable.

Lastly, in our opinion, the contractor would be involved in the events by not having acted diligently through opposition to the installation of an unknown material, provided it has a professional involvement in the choice of the materials to be used and, thus, will be liable for the damage caused by the professional liability pertaining thereto, always provided it is demonstrated that it was in charge and should have safeguarded the quality of the necessary materials.

THE PURCHASER'S OPPORTUNITY FOR RECOVERY, THROUGH LITIGATION OR OTHERWISE, FROM THE VENDOR'S ARCHITECT

The liability of this producer must also be analysed from the two perspectives indicated regarding the contractor.

1 GLU&C Art. 18

This rule establishes a type of objective liability: the producer is liable solely for having executed a job that implies a risk, regardless of whether or not it acted culpably.

However, this matter is not discussed sufficiently in our doctrine, nor is there case law in this regard. Given the absence of information, in our opinion, the application of this provision could be devised according to two interpretations:

(a) The producer is responsible for all damage caused by its activity, regardless of negligence. In this case, the architect must be liable despite his warning to the developer, since, although his actions fulfilled the requirement of competence and care usual in the activity of construction, objective liability does not admit a ground for exception once damage has occurred.

Since this solution may contradict the natural sense of equity, the court called upon to resolve the case might seek the following interpretation:

(b) The objective liability embodied here consists simply of a change in the burden of proof; the producer is liable if it causes damage to a third party by a negligent or intentional act; however, unlike the regime in the

Civil Code – where the plaintiff must prove the defendant's negligence or wilful misconduct –, the producer is presumed guilty and must prove that it acted with due care in order to be released from liability.

In this case, the architect's warning concerning the risks of using a new product without verifying its technical suitability releases him from liability, since the damage was caused by the actions of a third party, namely the developer, because it insisted on using the tiles, despite the warning. If the architect can provide evidence of his warning, he will not be liable.

2 Rules of the Civil Code on offences and quasi-offences (Arts 2314 ff.)

The degree of diligence required of a producer is the same as the professional liability in the GLU&C, so his warning releases him from guilt. The difference regarding the special regime is that here he is presumed innocent, so the plaintiff is the one who will have to prove the architect's negligence, and the architect will counter-attack by demonstrating his initial refusal to use the material.

If the architect had not given the warning, he would be liable pursuant to both regimes: the special one in the GLU&C Art. 18, and the general one in the Civil Code. The plaintiff will have to make a choice, as indicated in regard to the contractor, meaning that it will be more convenient to choose the GLU&C-regime because the evidentiary system favours it, save in the *in solidum* issue if it wishes to sue both producers at the same time.

Lastly, if the plaintiff chooses to resort to GLU&C Art. 18, it may, pursuant to sub-paragraph two of the rule, either sue the body corporate through which the producer acted or the producer directly.

THE INVOLVEMENT (IF ANY) OF INSURERS IN THE PURCHASER'S CLAIM FOR RECOVERY, INCLUDING THE METHOD OF RESOLVING THE MATTER

In the last decade, insurance companies have developed 'warranty insurance.' Insurers have become specialised and engaged exclusively in the matter, setting themselves apart from general and life insurance. Several insurance companies have come into the business which deal only with the provision of Warranty and Credit Insurance.

In general terms, Warranty Insurance, according to the regulations approved by the pertinent agencies, is a contract whereby the insurance company undertakes to pay the damages suffered by the policy beneficiary up to the amount insured, in the event of an insured risk.

The insurance company will always be entitled to claim against the policyholder to be reimbursed for the amounts paid to the beneficiary.

There are two types of warranty insurance which relate to the form of execution or enforcement, namely *immediately enforceable warranty insurance policies* and *prior adjustment warranty insurance policies*.

The former are policies where the beneficiary receives payment from the insurer on demand. In other words, there is no adjustment by the company for the damage caused by the insured risk. This type of policy meets the same purpose as 'bind-bonds' and performance bonds.

The latter are policies where, upon the beneficiary's request, the company directly quantifies the losses or damages in the periods specified in the policy itself, and, after such a quantification, orders immediate payment of compensation to the beneficiary.

Both policies – the immediately enforceable policy and the prior adjustment policy – are subject to a period of forfeiture during which the beneficiary must enforce it. After expiration of the period, the policy in question lapses and the insurer will not be liable, even if the damage was caused prior thereto.

Warranty insurance may cover risks in different stages of construction or real estate business: there are policies to warrant the seriousness of bids or tenders, correct investment of advances provided by the client, faithful and timely execution of the work and correct execution of the same.

In this latter category – warranty policies to secure correct execution of the work –, it is possible to distinguish, in turn, between policies covering risks during the period of execution and policies covering risks once the work has been completed. In other words, policies whose term ends upon hand-over of the work, and policies which begin at hand-over. These latter constitute policies that protect the client or purchaser concerning failures in the quality of the works.

In reality, the large majority of warranty policies issued by insurance companies are those referring to the period of construction and not to the post-construction stage.

Notwithstanding the foregoing, and according to the terms of the warranty insurance policies authorised by the *Superintendence of Securities and Insurance*, it is possible to issue warranty policies for hidden defects or defects in construction. This position has led to insurance companies issuing this type of policy, but only for prior adjustment insurance, meaning there is no immediate enforcement insurance existing in Chile for the post-construction period.

The terms for which post-construction policies are now issued run from one to two years, although renewal is possible. The reason why the issue of longer-term policies is not common is the high price of the premium.

In conclusion, although it is true that there has been a significant development in the country in recent years in the issue of warranty policies to secure correct execution of the work, this development has concentrated largely on policies covering the construction period, even though, as mentioned, the legal framework does provide for the development of post-construction policies that tend to protect the client or a purchaser from construction defects.

At the margin of such general insurance, a specific construction company has implemented an insurance system for its residential clients. This system consists of a warranty of the quality of construction for a period of three years from hand-over of a house or an apartment. The company has itself defined standards on quality. If, during that period, any defect in the work is revealed that goes beyond the limits contained in the standards, the company assumes the cost of repair. In order to insure performance of this obligation, the company has also contracted insurance with an insurance company so that the purchaser has the warranty that if the contractor does not voluntarily respond, its insurer will.

The system of warranty described is accompanied by an after-sales service that consists of the contractor advising the purchaser on maintenance of the work and assuming repair of flaws specific to the initial period of its use. The warranty therefore operates regarding defects that do not form part of the 'start-up.' The criteria to distinguish between the two types of defects arise from the standards set by the contractor, which are communicated to its clients at the time of purchase.

This system of warranty exists only for residential works; it constitutes a private response to a public problem. It is the only entity in the industrial sector of construction in Chile that offers it.

THE PROCEDURES BY WHICH THE ABOVE ARE RESOLVED, INCLUDING EXTRA-LEGAL, LEGAL AND INSURANCE PROCEDURES

This question is dealt with also in the *Introduction.*

1 Extra-legal procedures

It is very common for the parties to negotiate directly in order to reach an agreement on the amount of damages and the form of payment. The developer itself often assumes the cost of repair. In other cases, the purchaser remedies the defects and the developer later reimburses it for expenses incurred.

In large-scale works, such as apartment buildings and industrial construction, the parties usually resort to a third party as mediator. The mediator's role is to conduct dialogues between the parties in order for them to be able to reach a mutually satisfactory agreement.

2 Legal procedures

It is very common for the parties to include an arbitration clause in the purchase-and-sale agreement whereby any conflict arising between them in relation to the contract is subject to the hearing of an arbitrator-at-law (see *Introduction*).

If the parties do not choose to submit the dispute to an arbitrator, they may resort to the ordinary courts. Courts currently hear such conflicts in an ordinary court trial. However, the possibility does exist that disputes relative to construction matters will be processed more rapidly and expeditiously in the near future by means of a summary procedure, cf. *Introduction*.

3 Insurance procedures

The way in which insurance will take effect will depend on whether the policy is 'immediately enforceable' or 'after adjustment,' cf. above.

ANY MECHANISMS EXISTING FOR UTILISING THE FEEDBACK OR EXPERIENCE FROM THE CASE TO A BODY OR BODIES WHICH MAY MAKE USE OF IT

There are no mechanisms on information feedback in Chile regarding this type of controversy.

APPENDIX

Bill amending the GLU&C for the purpose of favouring a better quality in construction

A Bill is currently being processed that amends several aspects of the GLU&C, including post-construction liability. This Bill has been approved by the National Congress and was sent to the President of the Republic, who must decide whether he will make observations. If he does not, the Constitutional Court will make a decision on the Bill's conformity with the Constitution.

Below we transcribe article 18 of GLU&C – the article governing liability – according to the text approved by Congress, and then comment on how the situation would change for the proposed case.

1 Article 18 of GLU&C according to the text approved by Congress

The owner and first vendor of a construction will be liable for all damage and injuries arising from failures or defects of the same, either during the execution thereof or after completion of the same, notwithstanding the right thereof to file claim against whoever is liable for the failures or defects in construction that have caused the damage and injuries.

Designers will be liable for the errors they have made if such errors lead to damage or injuries.

Notwithstanding the stipulations in paragraph 3 of CC Art. 2003, contractors will be liable for failure, errors or defects in construction, including the works executed by sub-contractors and the use of defective materials or inputs, notwithstanding the legal actions that they may file in turn against suppliers, manufacturers and sub-contractors.

Bodies corporate will be jointly and severally liable together with the competent producer acting on behalf thereof as designer or contractor regarding such damages and injuries.

The owner [client] and first vendor shall be obligated to include a list in the public deed of purchase and sale containing a description of the designers and contractors to whom liability may be attributed pursuant to this article. In the case of bodies corporate, the list shall describe their legal representatives. The conditions offered in advertising shall be considered incorporated into the purchase-and-sale agreement. Final drawings and technical specifications as well as the Job Book referred to in article 143 shall be kept on file in the Municipal Works Office, available to interested parties.

The civil liability referred to in this article, in the case of bodies corporate that have been dissolved, shall be in force regarding whoever was the legal representative thereof on the date of execution of the agreement.

Actions to enforce liabilities as referred to in this article shall be lost through prescription five years after the date of final approval of the work by the Municipal Works Office.

2 Modifications of the rule currently in force

The most important modification is that the Bill establishes the liability of the client and first vendor. The objective is for this subject to be the one to respond to a purchaser who has suffered losses or injuries due to errors or defects in the works. It is a type of objective liability, i.e. the first vendor is liable for the sole fact that damage has occurred, regardless of whether this was due to negligence or wilful misconduct. As a result, the Bill contemplates the modern trend to expand the scope of application of the theory of objective liability in industrial activities in order to protect citizens who are in general, according to the traditional notion of subjective liability (for wilful misconduct or negligence), at a disadvantage regarding the businessman undertaking a risky activity.

Once the first vendor has given redress to the injured party, the first vendor may sue the party responsible for the defect having caused the damage, according to general principles.

Also established is the liability of designers for errors having caused damage or injury. This was already in the existing text, but the Bill is more generic when speaking of 'designers,' which includes architects, engineers and any other producer involved in the elaboration of the project.

In conjunction with designers, contractors are liable both for their errors and for those of sub-contractors, and also for the use of defective inputs and materials. The reference to CC Art. 2003 is made in order to clarify the point that there is liability for defects that may cause the ruin of the building; for defects of lesser gravity, there is liability according to this special rule. Additionally, as in the current rule, the producer may sue the perpetrators of the defect that caused the damage, once it has compensated the purchaser for loss suffered.

Sub-paragraph four elevates to the level of law a rule that is currently contained in the ordinance of GLU&C, which establishes joint liability between bodies corporate and a producer acting on their behalf. Paragraph six adds that the legal representative of a body corporate at the time of dissolution shall be liable for such body corporate if it has been dissolved. The objective is to come to terms with the problem of contractors who form a company for a certain job and, once this is completed, dissolve it so that the injured parties have no one against whom to direct an action for compensation.

Sub-paragraph five states the obligation of the first vendor to include a list in the deed of purchase and sale, indicating the designers and contractors who may be held liable according to this article. This is intended to facilitate action by a subsequent owner.

The final sub-paragraph provides a term of five years for prescription of actions derived from this article, in terms similar to those in the existing rule.

3 Answers to questions in the proposed case under the Bill

3.1 Can the purchaser (the Bank) receive redress for its damages by filing legal action against the vendor (the developer)?

The developer fulfils the role of client and first vendor and is subject to an objective standard of liability. It is therefore obliged to respond without any need for the purchaser to prove that its conduct – ordering the inclusion of the tiles – was negligent; it must respond by the mere fact that the damage occurred.

3.2 Can the purchaser be compensated for its losses by bringing an action against the contractor?

Because it is an objective liability, the contractor must respond, notwithstanding its right to sue the manufacturer of the tiles at a later date.

3.3 Can the purchaser be compensated for its losses by bringing an action against the architect?

(a) Because there is an objective liability, the architect must respond, despite having warned the developer of the risk of using a product unknown in the country.

(b) The architect is not liable since, by having opposed the use of the tiles and using them only once the real estate company insisted, it becomes an error of the latter, and objective liability cannot attribute the consequences of a harmful event to a person other than the one committing it.

The judge is the one to determine the scope and depth of application of the theory of the risk (objective liability) and specifically whether the producer may be released from liability, stating for the record his opposition, or whether under no circumstances a producer may adopt a course of conduct that can cause damage.

Type B–I

Alberto Ureta Alamos, Juan Carlos León Flores, Cristián Canales Palacios, Juan Claudio Molina and Gastón Escudero Poblete

The *Muñoz* v. *Camus et al.* case

1 The Plaintiff
The Plaintiff – a Sr. Ricardo Muños Valverde, an attorney by profession – purchased an apartment in a summer resort, from the developer Constructora Sol Ltda. on July 9th, 1980, in order to rent it out during the vacation months and enjoy it with his family during the rest of the year.

2 The Works
The construction was authorised by the Municipality of Viña del Mar on September 7, 1979 and final approval took place on April 22, 1980.

The sum paid by the Plaintiff for the apartment was equivalent at that time to some USD 6,500.

3 Procurement Method
The developer retained the services of several producers for the design and execution of the work; the method used is called 'By Management,' and means that the developer hires the services of architects, engineers, etc. and provides them with materials; the developer is the owner of the works from the beginning, and the intention is to make a profit from the sale of the same.

4 Insurance
Neither the purchaser nor the vendor had insurance for damage, nor for civil or any other type of liability.

5 Damage or Loss
The property suffered serious damage because of an earthquake that occurred on March 3, 1985. The country has a history of telluric movement, and earthquakes such as the one that damaged the building are not exceptional.

The underlying defect was inappropriate foundations.

10 Process after Discovery

The Plaintiff filed a claim before the 23rd Civil Court of Santiago against the architect, the structural engineer, the developer's legal representative, and the municipality that had granted the building permit and approved the works when completed.

He did not request that the damage be repaired but demanded compensation for the ruin of the building in regard to his apartment, compensation for the value of the apartment, for the profit he would no longer receive by not being able to rent it out as he did from the time of purchase to the event (consequential loss), and for 'moral damage' (*daño moral*).

11 Technical Reports

The process by which the lawsuit was heard envisages the possibility that the parties submit technical reports, or that the court on its own initiative requires such reports, among them proof of the events that took place.

The plaintiff commissioned soil surveys and other analyses at a university laboratory and paid for them.

The defendants submitted technical reports, prepared during construction, on the land's suitability for being built upon. They did not request or produce new reports once the lawsuit had begun.

20 Resolution

The issue was resolved judicially in regard to all defendants, except for the structural engineer, with whom the Plaintiff reached an agreement after the first instance hearing was concluded.

The first instance hearing of the lawsuit was before a civil court in Santiago. The parties were represented by their attorneys, substantiating their claims through writs, replies and rejoinders, and submitted the evidence they considered pertinent.

As the ruling overwhelmingly, though not entirely, favoured the plaintiff, the defendants appealed to the Court of Appeals of Santiago, who heard the appeal and claim that the judgment was technically void (except for the structural engineer who, as mentioned, had reached an agreement with the plaintiff). The Court of Appeals favoured the defendants, and the plaintiff then appealed to the Supreme Court on the grounds that the decision was technically and, alternatively, substantively incorrect.

21 Costs
The costs of the lawsuits were divided among both parties. The Supreme Court did not condemn the defendants to pay costs, because they did not totally lose their case.

22 Information
All parties had access to all information such as technical analyses, expert opinions, witnesses, data, etc.

23 Time
In the first instance, the case took six years and three months (from May 1985 to August 1991). In the second instance (in the Court of Appeals), the case took two years (from August 1991 to August 1993), and in the Supreme Court it took one year and 11 months (from August 1993 to July 1995).

24 Role of Experts
The courts accepted all expert reports submitted by the parties, and also made inspections of the building, accompanied by the parties with their experts, but did not request expert reports on their own account.

25 Outcome
As mentioned, the first court largely favoured the Plaintiff.

The Court of Appeals rejected all parts of the suit on account of the action being statute barred, thus rejecting the ruling by the lower court.

The highest court in the country rejected the first of the remedies mentioned above, but sustained the second, confirming the ruling in the first instance, which had, as mentioned, endorsed most of the claim. The Court ordered the defendants to pay compensation for consequential loss and loss of profits, plus adjustments and interest.

As mentioned, one of the defendants reached an agreement with the plaintiff by virtue of which he paid a certain amount in compensation.

30 Feedback
Any person may consult the complete file on the case with the first instance court, which is the one in charge of enforcing the ruling by the Supreme Court.

After a certain time, files are taken to a public agency called the Judicial Archive, which keeps files on all judicial causes heard in ordinary courts. There again, the file may be consulted by anyone.

Moreover, rulings in cases of some interest are published in several law reviews. The decision in the lawsuit commented upon here was published in the *Revista de Derecho y Jurisprudencia* (Law and Case Law Review).

Type B–II

The *Fourth Storey* case

1 The Plaintiff
The plaintiff was an individual who purchased a property for investment, intending to lease it out.

2 The Works
The completed building, located in Santiago de Chile, was a four-storey building with a converted loft, with office and residential accommodation.

The vendor had acquired it as a three-storey building to which he added a fourth floor and the loft.

3 Procurement Method
The Plaintiff entered into a purchase-and-sale agreement through which he acquired the entire building (including the new facilities built by the vendor).

4 Insurance
No insurance was involved.

5 Damage or Loss
The building purchased by the plaintiff suffered from serious defects on the fourth floor and the roof; the damage led to a decree by the Mayor's office (head of the municipal government) calling for demolition of the fourth floor and the roof.

The damage was detected by the plaintiff three years after execution of the purchase-and-sale agreement. The delay in discovering the defects was due to the fact that the vendor had made repairs which did not allow the purchaser to perceive the damage easily, and only the passing of time and use caused them to become apparent.

The plaintiff requested compensation for his loss, namely
- consequential damage, i.e. expenses incurred in the repair of the defective construction;
- damages for loss of profit, i.e. losses suffered because the property could not be leased;
- the costs of the expert assistance which the plaintiff had to obtain.

10 Process after Discovery
Once the damage had been detected, the plaintiff brought an action against the vendor.

11 Technical Report
The plaintiff requested expert reports at his own expense. These were produced before the action was brought. The defendant did not commission any technical reports.

20 Resolution
The conflict was resolved by a judicial ruling. This was an ordinary claim heard before the courts.

A claim was filed before the first instance court; the plaintiff appealed against its ruling to the Court of Appeals of Santiago, and, having obtained the ruling of the Court of Appeals, filed an appeal to set aside the judgment for an error of substance that was heard by the Supreme Court.

21 Costs
Each party paid its own costs.

22 Information
The parties had access to all the reports.

23 Time
The time taken was 17 months in the first instance, two years for the appeal and two years in the Supreme Court.

24 Role of Experts
Only the plaintiff used his own experts. The defendant did not contradict them because his defence was based on the prescription of the actions filed by the plaintiff.

25 Outcome

The conflict was resolved by a ruling of the Court of Appeals rejecting all parts of the suit because the plaintiff's right of action had been lost by prescription.

26 Feedback

Information on this case is available to professionals through publications containing the most typical decisions in decided cases.

---

[1] A. Alessandri, M. Somarriva, A. Vodanovic, 'Derecho Civil,' Vol.I, 5th Edition 1990, Editorial Conosur, p.117

[2] Arturo Alessandri, 'De la Responsabilidad Extra-contractual en el Derecho Civil Chileno' (Extra-contractual Liability in Chilean Civil Law), Vol.I, Ediar Editores Ltda., 2nd edition 1983, p. 11

[3] Santiago Court of Appeals, August 19, 1884 Court Gazette, 1884, No. 2030, p.1244

[4] Ibid.

[5] Arturo Alessandri, 'De la Compraventa y de la Promesa de Venta' (The Purchase and the Sale Promise), Vol.II, Soc. Imprenta Litografica, Barcelona 1917, p.264

[6] Santiago Court of Appeals, June 21, 1882, Gazette of the Courts, 1882, No. 1,257; p.731

[7] Supreme Court, January 13, 1994. Gazette of the Courts, first semester, No. 5, p.74

[8] Ricardo Sandoval L., 'Manual de Derecho Comercial' (Manual of Commercial Law), Vol.I, Ed. Juridica de Chile, 3rd edition 1990, pp.130-1

[9] Ibid., p.85

[10] Ibid., p.88

[11] Revista de Derecho y Jurisprudencia (Law and Case Law Review), Vol. XLVII, 2nd part, sect.1, p.452

[12] Revista de Derecho y Jurisprudencia (Law and Case law Review), Vol.27, sect 1, p.323

[13] Ibid., Vol.13, Sect. 1., p.110

[14] Arturo Alessandri, 'De la responsabilidad Extra-contractual en el Derecho Civil Chileno' (Extra-Contractual Liability in Chilean Civil Law), Vol.I, p.377

[15] Ibid., p.364

[16] Court of Appeals of Iquique, July 6, 1918. Gazette of the Courts. July-August no. 308, p.957

[17] Arturo Alessandri, op. cit., p.368

[18] Supreme Court, January 13, 1994. Gazette of the Courts, first semester, No. 5, p.74

China (People's Republic)

Type A

Deepak Bajaj with John Twyford

THE PURCHASER'S OPPORTUNITY FOR RECOVERY THROUGH LITIGATION OR OTHERWISE, FROM THE VENDOR

At the outset, it is assumed that the developer is an enterprise within the principles set out in the General Principles of Civil Law (1987) p.225 of the 1983-1986 Book (hereafter referred to as the *Civil Code* or CC), Art. 41.

Depending on the terms of the contract for the sale of the building, there is the potential for the purchaser to recover losses from the developer. This would require an appropriate provision in the contract whereby the developer incurred a contractual obligation in respect of the economic viability of the building, CC Art. 111. Under CC Art. 35, there is a period of limitation of two years within which action must be commenced, and this requirement appears to have been met.

THE PURCHASER'S OPPORTUNITY FOR RECOVERY THROUGH LITIGATION OR OTHERWISE, FROM THE VENDOR'S CONTRACTOR

The purchaser would not be able to recover in contract from the vendor's contractor, as CC Art. 111 clearly refers to the rights of a 'party' to the contract, and the purchaser is not a party to that contract.

Case Studies in Post-Construction Liability and Insurance, edited by Anthony Lavers. Published in 1999 by E & FN Spon, 11 New Fetter Lane, London EC4P 4EE, UK. ISBN: 0 419 24570 7

It is possible that the contractor would be liable under CC Art. 122 on the basis of product liability. This would involve the concept that he became the seller of the defective material by incorporating it into the building. This liability is by no means certain, as Art. 2 of the Civil Procedure Law (1982) found at p.259 of the 1979-1982 Book (hereafter referred to as the *Procedure Code* or PC) requires the People's Courts to 'distinguish right from wrong', and it might be argued that the contractor had no involvement in the selection of the defective material. Had the contractor known about the defective material through technical literature, but used it, the case against him would have been strong.

THE PURCHASER'S OPPORTUNITY FOR RECOVERY, THROUGH LITIGATION OR OTHERWISE, FROM THE VENDOR'S ARCHITECT

It seems that the architect would be exempt from liability, as CC Section 3 does not contemplate a wide basis of liability for negligence. The architect might, however, be liable under CC Art. 130 as having 'jointly infringed' (with the developer and the contractor) the rights of the purchaser.

THE INVOLVEMENT (IF ANY) OF INSURERS IN THE PURCHASER'S CLAIM FOR RECOVERY, INCLUDING THE METHOD OF RESOLVING THE MATTER

Nothing in the CC suggests how an insurer might become involved, though presumably the tile manufacturer would carry product liability insurance, but the PC does not suggest that an insurer could be made a party to any proceedings.

It is worth noting that the tile manufacturer is an overseas company, since special provisions relate to foreigners (CC Chapter 3).

THE PROCEDURES BY WHICH THE ABOVE ARE RESOLVED, INCLUDING EXTRA-LEGAL, LEGAL AND INSURANCE PROCEDURES

If the dispute involves only citizens of China or enterprises situated in China - see above -, the matter will be dealt with in the People's Courts, under the procedure set out in the PC. The PC emphasises the avenue of conciliation, both before and after the hearing.

The court system is a four-tier system, and the level of jurisdiction that will hear the matter will depend on the nature of the dispute and on the identity of the parties.

The mode of trial contemplated by the PC appears to be inquisitorial, with the tribunal having the power to appoint experts.

PC Chapter XX recognises arbitration as a means of solving disputes involving 'foreign economic' interests.

ANY MECHANISMS EXISTING FOR UTILISING THE FEED-BACK OR EXPERIENCE FROM THE CASE TO A BODY OR BODIES WHICH MAY MAKE USE OF IT

There is no information in either code that would suggest an answer to this question.

Type B–I

Deepak Bajaj with Zhong Li, Ming Rong Shi, Rui Zhang

The *Shanghai Jing Pu Mansion* case

1 The Plaintiff
The plaintiff was Shanghai Hai Property Development Co. Ltd., a joint venture development company in the public sector. This company often commissioned construction works.

2 The Works
The project was an office building in Shanghai City's Central Business District location commissioned by the plaintiff, and completed in 1995.

3 Procurement method
Procurement was by a traditional open tendering system.
Contractor: Nan Tong Construction Corporation.

Cement supplier: The client nominated the cement supplier (Shanghai Zhong Shan Industrial Corporation), who supplied cement produced by the Qu Xi Factory.

4 Insurance
No insurance has been mentioned in this case.

5 Damage or Loss
Construction by the contractor, Nan Tong Construction Corporation was in progress when it was found that the cement used in the structure between levels 11 and 14 was below the 'GB Standard' (National Standard System of the People's Republic of China).

The error, discovered by a random quality check conducted by a government agency at the completion of level 14, was in the material, its testing and site supervision.

The loss for which compensation was claimed was the cost of demolishing (CNY 2.11 million) and rebuilding the defective levels, in all CNY 6.62 million.

10 Process after Discovery
The client initiated the claim, lodged with the Municipal Court. The first defendant was the material supplier, the second, the manufacturer, and the third, the contractor.

11 Technical Reports
Neither the client - who inferred that the contractor was presumed to test material arriving on the construction site - nor the defendants had reports made.

20 Resolution
The first court session was held on December 14, 1995.

The plaintiff claimed that although the plaintiff had nominated the supplier himself, the cement was always delivered directly to the construction site and, presumably, there tested by the contractor. The purchase order had been placed only after the test of the first batch of 20 tons was found to be up to standard. Therefore, the plaintiff argued, the onus of cement quality assurance was upon the contractor. Furthermore, the quality certificates had been sent directly from the supplier to the contractor, and the plaintiff had never seen them before.

The supplier claimed that the manufacturer had shown him quality certificates, and that the cement was dispatched to the site according to schedule. The damage was caused entirely by the low quality of the cement manufactured by the Qu Xi Factory, which therefore, he maintained, should bear the whole cost.

The manufacturer argued that the evidence produced by the plaintiff was not strong enough to prove that all the cement used on the site was from his factory. Furthermore, the contractor did not test the quality of the cement, so the defect, he claimed, was caused by negligence on site.

The contractor argued that it was evident that the low quality cement was the reason for the damage. The cement supplier had been nominated by the client, and there were approval stamps both from the supplier and from the manufacturer on the quality certificates. Furthermore, when the contractor discovered the sub-standard quality, he had immediately informed the client, suggesting that work be halted. However, the client had refused his request to discontinue construction; written evidence of this refusal was produced in court.

21 Costs
The conflict resolution costs are not known.

22 Information
All information was available to all parties.

23 Time
Three months elapsed from the first court trial to the Government Agency's intervention (below).

24 Role of Experts
All parties produced and presented their own evidence.

25 Outcome
The client agreed to a non-litigation resolution, but the first defendant, the supplier, declined to accept this approach.

The case caused a fervent national debate with a dramatic conclusion when the highest government authority relevant to the building industry in Shanghai, the *Shanghai Building Industry Administrative Office*, eventually laid down the rule, as follows:

The manufacturer and the supplier should take the major responsibility, which rendered each of them liable to pay 35 per cent of the damage, while the client and the contractor should share the remaining 30 per cent equally.

In addition, the Office suggested that both the supplier's licence for cement wholesaling and the manufacturer's licence for producing cement be revoked. No court action relevant to this suggestion has been reported.

30 Feedback
All facts are public.

Denmark

Jens Jordahn

Type A

THE PURCHASER'S OPPORTUNITY FOR RECOVERY, THROUGH LITIGATION OR OTHERWISE, FROM THE VENDOR

(a) As a result of the legal principle according to which a vendor, whether negligent or not, must pay damages (or accept a *pro rata* reduction), the purchaser will recover (most of) its loss from the vendor. This outcome is all the more evident as the developer, if the case went to trial, would also be found negligent: Acting against the advice of a designer will — at least for a developer — be considered negligent.

It is of no importance that the purchaser did not inspect the building, as he is under no obligation to do so. However, if the developing cracks could be seen by anyone without being an expert, an obligation to make inquiries or further inspections would lie on the purchaser.

It could also be argued that the vendor could or should inspect the building before selling it.

The vendor's liability is not affected by the building research institute's being aware of the problem.

(b) The owner is under an obligation to repair as soon as the experts appointed by the court have rendered their opinion, and no further examination is needed. If he fails to carry out repairs, the owner will not be reimbursed for that part of the remedial costs which is attributable to the repair being delayed.

Case Studies in Post-Construction Liability and Insurance, edited by Anthony Lavers. Published in 1999 by E & FN Spon, 11 New Fetter Lane, London EC4P 4EE, UK. ISBN: 0 419 24570 7

THE PURCHASER'S OPPORTUNITY FOR RECOVERY, THROUGH LITIGATION OR OTHERWISE, FROM THE VENDOR'S CONTRACTOR

The contractor performed his work according to specifications. He did not know – and nothing indicates that he ought to have known – the risk involved in using the particular product for the cladding tiles. According to legal practice, the contractor only has to object to the design if it is evident to him that work according to the specifications will lead to damage. The contractor therefore has no liability for the purchaser's loss.

Furthermore, it is a rule that 'the contractor shall not be liable for operational losses, loss of profit or other indirect losses' (AB92 Cl. 35, subs. 02[i]). This might not be accepted by the Courts in cases of gross negligence, but that is not the case here.

THE PURCHASER'S OPPORTUNITY FOR RECOVERY, THROUGH LITIGATION OR OTHERWISE, FROM THE VENDOR'S ARCHITECT

In 1973 (UfR1973.675H, see also UfR1973.322B, (below)), the Supreme Court held that for an architect it is not acting within the 'state of the art' to accept a new material in a construction, unless either sufficient documentation for the suitability of the product is obtained or the Client is informed that it is a new product and that no documentation of its suitability exists.

But, as for a contractor, it is a general rule that a 'consultant shall not be liable for working deficits, loss of profits or other indirect losses' (ABR89, Cl.6.2.4[1]). The wording of this clause – which has to be expressly accepted by both parties to be binding – speaks for itself, and it might not be accepted by the Courts in cases of gross negligence.

In the case under scrutiny, there are further points to be taken into account concerning the architect's liability to the purchaser:

(a) It is undisputed that the architect objected to the use of the special product on which the developer insisted.

It is open to discussion whether the architect should have declined liability in writing and, by failing to do so, gave the developer the impression that the product would be acceptable, thereby assuming liability. However, here a developer refused to accept the advice of his architect.

Thus, it seems more proper that the developer bears the risk of the unknown effect of the new material.

(b) The architect has no contractual obligation towards the purchaser. Therefore, no contractual liability can be established.

It could be argued that the architect is liable in tort to the purchaser: Knowing that he is rendering his service to a developer who is most likely to sell the building and use the architect's name, the architect must also act on behalf of such a purchaser. Therefore, he could be liable for not making specific reservations.

It has been the opinion in Denmark that the architect's liability in such cases cannot exceed the liability of the owner. However, in 1995 (UfR1995.484H, below) the Supreme Court awarded costs in tort in a situation where 'manifest professional errors were made regarding essential conditions concerning the building'. However, the defects caused by the cladding product cannot be established as the result of a fault of this kind, and the defects are of no importance to the structure and other parts of the building.

It is, therefore, highly unlikely that the architect would be considered liable for the loss. Even if he was, he would only be liable jointly with the developer, and, between the two, the developer would have to compensate the architect.

THE INVOLVEMENT (IF ANY) OF INSURERS IN THE PURCHASER'S CLAIM FOR RECOVERY, INCLUDING THE METHOD OF RESOLVING THE MATTER

Under Danish law, normally no owner's insurance will cover this kind of loss.

THE PROCEDURES BY WHICH THE ABOVE ARE RESOLVED, INCLUDING EXTRA-LEGAL, LEGAL AND INSURANCE PROCEDURES

(a) As the cause of the damage is unknown, the purchaser would notify the developer, who would notify the architect and the contractor.

If, at the time of notification, the developer knew that the building research institute was aware of the problem, he may decide to notify only the architect and not the contractor (unless, of course, bad workmanship is likely to be the reason for the damage).

(b) Settlement discussions between the parties would be initiated.

(c) Neutral experts (normally court-appointed) would examine the damage, before or after initiation of legal proceedings.

(d) Settlement discussions would take place when the experts had rendered their opinions.

(e) Litigation, arbitration, or both would be initiated. As it is unlikely that the sales contract with the bank would contain an arbitration clause, litigation between the vendor and the bank would be referred to the ordinary courts, while the case between the vendor and the consultant and the contractor would be referred to arbitration.

(f) After exchange of the parties' written pleas (normally two from each side), and, as the case may be, additional reports from the experts, the hearing in court or before arbitrators would take place; this would include presentation of the case, hearing of the parties, of witnesses and of experts. The hearing would be concluded after oral pleadings by the parties' lawyers.

(g) The court would suggest a decision or settlement to be considered by the parties.

(h) A judgment from a lower court can be appealed to a High Court, but only to the Supreme Court by special leave, which is only rarely given.

CONCLUSION

The purchaser will be fully compensated by the vendor for the cost of rectifying the defects and his loss in not being able to let the unit, as the vendor acted negligently.

If the vendor acted in good faith, the compensation would be calculated to take into account the difference in the price of the building at the time of transfer without the defect and with the defect (a *pro rata* reduction). However, in minor cases, i.e. where the costs to be recovered are less than some 10 per cent of the transfer sum (depending on the type of defect and the time expired after transfer), the purchaser will not be reimbursed. It is worth noting that for a developer the level of good faith is lower than for a non-professional vendor.

ANY MECHANISMS EXISTING FOR UTILISING THE FEED-BACK OR EXPERIENCE FROM THE CASE TO A BODY OR BODIES WHICH MAY MAKE USE OF IT

(a) If the case goes to the Supreme Court, the result will normally be published in an official weekly publication *Ugeskrift for Retsvæsen* (UfR) the Danish 'Legal Gazette'. If the case goes to a lower court only, or to the High Court, it will be published in UfR only if it is found to be of some general interest.

If the case is decided by The Building and Construction Arbitration Court in Copenhagen, which is instituted in AB92, Cl.47 and ABR89, Cl.9.0.1, the decision of the arbitrators, and sometimes the settlement proposals, will be published in a special publication 'Decisions concerning Real Estate', *Kendelser om fast Ejendom*, or KfE.

Type B-I

The *KfE (1986.21)* case

1 The Plaintiff
The plaintiff was one of the smaller municipalities. It often commissioned construction works, but had no in-house expertise.

2 The Works
The building was a public school, built 1976–77, which was constructed with a flat roof made of roof coffers. Flat roof coffers were often and normally used at that time.

3 Procurement Method
The contracts were traditional separate trades.

4 Insurance
Whether the Client had insurance is unknown. The designers carried the normal professional liability insurance against negligence. The Contractor carried no insurance for damage to its own works.

5 Damage or Loss

Some roof coffers were damaged by rain before they were built in, and should have been replaced. This was discovered in August 1980, some two years after hand-over.

The Client claimed some DKK 2 million.

The contractor had been instructed by the supplier to protect the coffers against rain, from the time they were delivered to site until the last layer of roofing felt had been laid. It also appeared from the minutes of site meetings that the engineer had several times pointed out to the contractor that he should protect the materials from rain until the final roofing felt was laid; the engineer had furthermore instructed the contractor to drill holes in the coffers to drain away rainwater. It could be argued that the engineer, by giving specific instructions during construction, intervened in a way, which could be seen as new design, for which the engineer would be liable. This, however, was not argued in the case.

The coffers were partially repaired and repainted before hand-over.

No problems were observed (and neither notification nor reservation was presented, nor were deductions made from certificates) at the time of hand-over, nor at the expiry of the one-year defects liability period, nor when the performance bond was released.

According to the findings of the arbitration, the error was to be found in workmanship.

The damage was discovered by the Client (a public official).

10 Process after Discovery

The claim was initiated by the Client. Notification was given to the contractor, the engineer and the architect (it is not known whether the plaintiff's insurer was informed of the claim).

The designers informed their insurer of the claim.

11 Technical reports

The plaintiff requested and paid for a technical report written by a technological institute.

Also the defendants had a technical report made, and the professional liability insurer, using its in-house engineers and architects, had its own technical reports made to evaluate the claim, as is normal. These additional reports were paid for by the insurer.

Furthermore, the Client requested another expert to give his opinion.

Finally, the technical members of the Arbitration Board (below) examined the roof and prepared a short factual statement. There was no need for court-appointed, neutral experts, as the technical problem and the solutions were evident to the technical arbitrators.

20 Resolution

The matter was resolved by arbitration at the Building and Construction Arbitration Court in Copenhagen. Prior to the arbitration procedures, settlement negotiations had taken place, as is normally the case.

Alternative conflict resolution procedures such as mediation, conciliation or adjudication do not normally take place in Denmark.

The Rules of Procedure for the Building and Construction Arbitration Court are similar to the rules applied by the ordinary courts. According to these rules (Arts 3, 10 and 11), each party can submit two pleas to the court. They shall state their opinion and refer to the facts of the case, the documents they rely on, other evidence to be produced and the law. The parties to this kind of arbitration are normally represented by lawyers. At the oral hearing, witnesses are heard (they are not sworn; if this is needed, the case must be taken to court). After the presentation of the case and the hearing of the parties and the witnesses, lawyers make their final oral presentation to the court. Each party has the opportunity to address the arbitrators twice. Following this, the arbitration court will normally present its settlement proposal which will (always) have the same outcome as the eventual decision. If the parties do not accept the settlement proposal, a decision is rendered.

The decision will be accepted by the ordinary courts if the party against whom judgment is given does not fulfil its obligations.

The party who wins the case will normally be awarded costs, not according to its actual costs, but according to a scale which for larger cases will be 3 per cent of the disputed amount.

21 Costs

Own (internal) costs and costs for in-house technical expertise are not reimbursed. The party against whom judgment is given will have to pay the costs of external experts.

22 Information

All parties to the conflict had access to all studies, research and data concerning the issue needed for their pleading.

23 Time

The Client's technical report was issued in December 1980, and additional reports were made in 1981 and 1982.

Arbitration procedures were initiated in February 1983, as settlement discussions had not led to any solution. The examination by the technical arbitrators took place in November 1983.

The pleadings from the parties were exchanged in 1983–84, and the hearing before the Arbitration Board took place in the winter of 1985/86, with a decision handed down on March 14, 1986.

24 Role of Experts
See above.

25 Outcome
The arbitrators' decision was as follows:

The contractor was found responsible for the damage, as he had not complied with the requirement to protect the coffers during construction, even after the engineer's notification.

The engineer who supervised the works and had instructed the contractor to be more meticulous might, it was found, have had to perform more examinations before hand-over and release of the performance bond. However, as the contractor was indisputably responsible, the arbitrators did not find sufficient reason to hold the supervising engineer responsible jointly with the contractor.

The Client was awarded DKK 220,000, with interest from June 1985 (when the claim was presented), and awarded costs of DKK 30,000.

The Client had to pay costs to the architect and the engineer with DKK 25,000 to each party.

The costs to the arbitrators were to be split evenly between the Client and the contractor.

There are no details of any insurance settlement as no insurance company was involved on the contractor's side.

30 Feedback
The case, with the decision, is published anonymously (Rules of Procedure for the Building and Construction Arbitration Court Art. 26) in *Kendelser om fast ejendom* ('Decisions on Real Estate'). However, the Danish construction market is so small that the job involved will often be known to persons working in the industry.

Type B–II

The *Phoenix* case

1 The Plaintiff
The plaintiff was an association of private owners of flats in a residential building; the Association did not normally commission construction works.

2 The Works
The building was a block of residential flats for several households. The roof was insulated by expanded polystyrene slabs protected by a layer of felt. When the slabs were in place, roofing felt tape was used for proofing according to the normal procedure at the time of construction. The roofing technology was innovative.

3 Procurement method
The block was built by a vendor consortium. Roofing was undertaken by a specialist contractor who issued a warranty according to which he accepted liability for defects, both in the quality of the felt and in workmanship. However, the warranty covered only repair of the felt.

4 Insurance
There is no information on insurance.

5 Damage or Loss
The Client discovered leaks in the roof in 1982, i.e. some two years after hand-over (under Danish law, there is no distinction between provisional or preliminary and final hand-over). The main reason for the problems was described as follows by court-appointed experts:

> The polystyrene slabs are exposed to considerable movement due to changes in temperature. The roofing felt which covers the slab butts will be torn due to these movements. Another (minor) cause is various arrangements in 'valleys' between the slabs.

According to the technical experts, the cause of the leaks was the use of polystyrene, and this could objectively be stated to be an error. However, at the time of design and construction, the use of such polystyrene products was in conformity with the state of the art. The quality (or lack of quality) of the material was in conformity with the contract. Thus, the felt contractor could not be held liable, either for the leaks in the roofing felt or for the movements in the polystyrene slabs delivered by the main contractor. Nor would a designer be held liable for the defect, as the design was, as mentioned, in accordance with the state of the art at the time.

10 Process after Discovery

The plaintiff notified the defendant in good time, and during the following years the roofing contractor tried – as was his obligation under the warranty – to remedy the defects, but with little or no effect. Private settlement negotiations did take place but were unsuccessful (non-litigation conflict resolution procedures, e.g. arbitration, mediation, conciliation or adjudication were not used; they are not normally used in Denmark).

The plaintiff then, in April 1989, initiated litigation against the contractor in a High Court, claiming costs to replace the roof and the costs to remedy ensuing damage to the building (DKK 423,000).

Two legal pleas from each party were, as is normal in High Court cases, exchanged according to The Danish Administration of Justice Act, Ss 348, 351 and 353 (the act is now known as Act No. 752 of 15 August 1996).

The expert appointed by the High Court rendered his opinion, and additional opinions were also collected. In 1992, the oral hearing took place, where the parties, witnesses and experts were heard and the lawyers made their final oral presentations to the Court. Each party had the opportunity to address the Court twice. The High Court rendered its verdict on December 18, 1992.

The judgment was appealed to the Supreme Court where the parties presented additional material. The Supreme Court gave its judgment on May 5, 1994.

11 Technical reports

The plaintiff did not have a technical report made, but requested that a court-appointed neutral expert deliver a report. This expert was a specialist in this field. It is unknown whether the contractor commissioned a technical report. The plaintiff had to pay the initial expenses for the report.

20 Resolution

The High Court held the contractor partly liable. The Supreme Court's verdict was in favour of the roofing contractor.

21 Costs

The High Court awarded the plaintiff lawyer's costs of DKK 30,000, plus the costs for the expert report, DKK 12,333. The Supreme Court awarded the roofing contractor costs of DKK 50,000 against the plaintiff.

22 Information

Both parties to the conflict had access to the neutral reports. However, if a party requests its own technical investigation, which is not presented to the Court or the Arbitration Board, the other party will not have access to such a report.

23 Time

The damage was, as mentioned, discovered in 1982. Pleadings and expert opinions were exchanged from 1989 through the spring of 1992.

The hearing of the case before the High Court, which delivered its verdict on December 18, took place in the autumn of 1992.

The appeal to the Supreme Court was made in January 1993.

Additional information was provided to the Supreme Court, where the case was judged in April and May 1994, with a verdict on May 5.

24 Role of experts

As mentioned, the plaintiff did not use its own experts but asked the High Court to appoint experts.

25 Outcome

The High Court found that the leaks in the roofing felt were covered under the warranty and ordered the contractor to pay damages for this part of the costs, estimated by the Court at DKK 200,000. However, the Court gave judgment in favour of the contractor for the rest of the costs, as his work had been performed according to the contract and according to the state of the art.

The Supreme Court found that the wording of the warranty did not include leaks in the roofing felt due to causes unconnected with the roofing felt itself or its installation. It therefore released the contractor from all claims.

30 Feedback

The case is published in the periodical *Ugeskrift for Retsvæsen*, the Danish 'Legal Gazette'.

---

[i] AB92: 'General Conditions for the provision of works and supplies within building and engineering' and ABR89: 'General Conditions for consulting services' are Agreed Documents; to a large extent they reflect what would be the rules according to Danish law.

England, Wales and Northern Ireland

Type A

Nabarro Nathanson (Solicitors)

THE PURCHASER'S OPPORTUNITY FOR RECOVERY, THROUGH LITIGATION OR OTHERWISE, FROM THE VENDOR

The primary legal relationship between the investment bank (the purchaser) and the developer (the vendor) is a contractual one based upon the sale and purchase contract.

The first question to address is whether the contract included any express warranty as to the construction of the building or that it was free from defects. An express warranty of this kind would give the basis for a claim for breach of contract.

In the absence of such, the issue is whether there is an implied warranty as to the condition of the building or obliging the vendor to notify the purchaser of any defects. The basic rules of contract preclude the court from interfering in a contract between parties of equal bargaining power simply on the basis that one party made a bad bargain. The maxim *caveat emptor* (let the buyer beware) forms the basis of all such contracts. There is no general obligation on the vendor to inform the purchaser of any defects: *Turner v Green* [1895] 2 CH 125; *Greenhalgh v Brindley* [1901] 2 CH 324; *Sheppard v Croft* [1911] 1 CH 521.

The existence of the maxim *caveat emptor* means that the onus falls on the purchaser to carry out surveys and inspections prior to purchase, to ensure that the building is free from defects or that he is paying a price commensurate with the existence of defects. The purchaser failed to carry out such checks here, and has no recourse for its own failure.

Case Studies in Post-Construction Liability and Insurance, edited by Anthony Lavers. Published in 1999 by E & FN Spon, 11 New Fetter Lane, London EC4P 4EE, UK. ISBN: 0 419 24570 7

In the absence of any contractual claim, it is necessary to look at whether any tortious claim exists. The purchaser's alternative approach would be a claim for negligence. The fact that the purchaser also has a contractual relationship with the vendor does not generally preclude this (*Holt v Payne Skillington & Another* (1995) 77 BLR 51), unless there is an express contractual provision doing so. It is assumed no such provision exists here.

Generally, a vendor would not have any duty of care in tort to a purchaser in respect of the building, but in this instance the vendor was also the developer and developed the scheme with a view to disposal, i.e. it had purchasers in its reasonable contemplation in carrying out the development (*St. Martin's Property Corporation Ltd v Sir Robert McAlpine & Sons Ltd* [1992] 63 BLR 1). Whilst the vendor did not have any construction professionals on its staff, it did have significant development experience.

In judging whether the vendor has breached its duty of care, it will be judged against the standards of a reasonable developer (*Bolam v Friern Hospital Management Committee* [1957] 1 WLR 582; *Maynard v West Midlands Regional Area Health Authority* [1984] 1 WLR 634). It is arguable that a reasonable developer would either have knowledge of materials which it expressly requires to be used or would listen to an architect's advice, although in this instance the architect did not object to the cladding tiles on the basis of potential defects but simply on the basis of his lack of knowledge concerning them. The questions will turn on where responsibility for the choice ultimately lies, and there may be contributory negligence arguments between vendor and architect on this point (below).

If a duty of care and breach are established here, we have to turn to the question of losses recoverable. These are limited to the cost of rectifying physical damage to other property, for physical damage or rectifying any imminent danger to health or safety or both (*D & F Estates v Church Commissioners* [1989] AC 177; *Murphy v Brentwood District Council* [1991] 1 AC 398). In this instance, there are not stated to be any losses in these categories. The only extensions to this will be if the court can identify a special relationship between the parties (*Henderson v Merrett Syndicates Ltd* [1995] 2 AC 145; *White v Jones* [1995] 1 All E.R. 691). The authorities for this proposition are not construction cases, but the principles may apply although there is as yet no direct case authority. The general principle, however, appears to require a specific individual actually in contemplation to suffer loss, i.e. by analogy not just a purchaser but the specific purchaser. This test would not be satisfied here.

THE PURCHASER'S OPPORTUNITY FOR RECOVERY THROUGH LITIGATION OR OTHERWISE, FROM THE VENDOR'S CONTRACTOR

The purchaser cannot claim in contract against the vendor's contractor in the absence of a direct contractual relationship. There is no mention here of the existence of any direct collateral contract between the purchaser and the vendor's contractor, and hence no basis for a contractual claim exists.

The purchaser may have a claim in tort against the vendor's contractor. A contractor has a duty of care to parties in its reasonable contemplation not to cause physical damage to other property or injury as a result of defective work, and will owe such a duty to future owners of the building.

The question is then whether the contractor has breached his duty of care. Unless the building contract incorporates design responsibilities, the contractor's duty will be to construct in a workmanlike manner the building as specified to him, i.e. including the cladding tiles. He will be judged against the standard of an ordinary skilled contractor (*Bolam v Friern Hospital Management Committee* [1957] 1 WLR 582; *Maynard v West Midlands Regional Area Health Authority* [1984] 1 WLR 634). The contractor's only liability would be if he was aware of the problems with the cladding material and failed to warn the vendor, though there is no evidence that this was the case, since the material did not become generally known to be problematic until several years later and he had no duty to make enquiries about the material.

Accordingly, the purchaser is unlikely to be able to recover from the vendor's contractor in either contract or tort.

THE PURCHASER'S OPPORTUNITY FOR RECOVERY, THROUGH LITIGATION OR OTHERWISE, FROM THE VENDOR'S ARCHITECT

As with the vendor's contractor, the purchaser cannot pursue a claim in contract against the vendor's architect in the absence of a direct contractual relationship which is assumed not to exist here.

Once again, the purchaser may have a claim in tort. A professional such as the architect has a duty of care to parties within its reasonable contemplation not to cause physical damage to other property or injury as a result of defective design.

In this instance, the architect has design responsibilities for the design and specification of the building and will be judged against the standard of the ordinary skills of his speciality (*Bolam v Friern Hospital Management Committee* [1957] 1 WLR 582; *Maynard v West Midlands Regional Area Health Authority* [1984] 1 WLR 634). An architect will also generally have within its contemplation future owners and users of the building. Two factors will, however, be relevant: firstly that the developer in effect specified the cladding tiles and the extent of the architect's duty to research these, and secondly that the architect's knowledge is judged against the state of the art knowledge at that time, since, although the research institute had suggested unsuitability, the information was not widely circulated until later. Accordingly, the architect may have defences.

The involvement of the developer in the specification of these tiles gives the architect a possible defence of contributory negligence, i.e. that the developer should share in responsibility. The question will be whether the architect made sufficient enquiries concerning the material proposed by the vendor. The state of the art will be judged against the reasonable knowledge of an architect at the date that the cladding was specified and built in, and the enquiries which a reasonable architect would have made in such circumstances. Given that the journals had not yet published articles on the subject, it may be hard to establish liability on the part of architect.

The next aspect of a claim in tort is to look at the damages sought. As mentioned previously, recovery in tort will be limited to the cost of rectifying physical damage to other property, for physical injury to persons or for rectifying any imminent danger to health and safety, or several of these (*D & F Estates v Church Commissioners* [1989] AC 177; *Murphy v Brentwood District Council* [1991] 1 AC 398). From the information available here, the tile cracking is not causing any damage to other property or other parts of the property, nor is there a safety issue, and accordingly neither the cost of re-cladding nor the loss of rent will be recoverable, since there is no special relationship between vendor's architect and purchaser.

Accordingly, a claim against the vendor's architect cannot be made in contract, and whilst a claim may be available in tort, subject to possible defences, in the absence of establishing a special relationship, it does not permit recovery of losses of the nature suffered.

THE INVOLVEMENT (IF ANY) OF INSURERS IN THE PUR-CHASER'S CLAIM FOR RECOVERY, INCLUDING THE METHOD OF RESOLVING THE MATTER

It is unlikely that a vendor would be insured for breaches of contract (if any) or for any tortious duties to the purchaser. The contractor may have some product guarantee type insurance for workmanship or professional indemnity insurance for any design deficiencies in its design, or both. Since the contractor appears to have no liability, his insurance is irrelevant.

Architects are likely to have professional indemnity insurance for any claims against them arising from deficiencies in their services. There is no obligation placed upon architects by the RIBA or by any other body to have such insurance or to specify a level, although their contractual conditions of appointment with the vendor may incorporate such obligations. Generally, however, architects do have professional indemnity insurance, at some level.

Assuming relevant insurers accept that a claim falls within the policy, they might defend the claim on behalf of and in the name of the insured. Otherwise the claim would be dealt with as if uninsured (below).

THE PROCEDURES BY WHICH THE ABOVE ARE RESOLVED, INCLUDING EXTRA-LEGAL, LEGAL AND INSURANCE PROCEDURES

Prior to commencing formal legal proceedings it is becoming increasingly common to attempt to resolve disputes by alternative dispute resolution, such as mediation. Mediation is a consensus-based procedure, using a qualified mediator to attempt to facilitate agreement between the parties.

If a contract includes a provision that any disputes are to be referred to arbitration, then this will prevail and only if both parties consent be dealt with in the courts.

Otherwise, claims in contract and all claims in tort must be pursued in the courts, the exact level of the court depending on the amount in issue: above a certain level, cases are dealt with by the High Court, by a specialist section of the court which deals with construction and technical disputes.

Allocation of the costs of arbitration and litigation is at the discretion of the arbitrator or judge, but usually the successful party can expect to recover between 66 and 75 per cent of its legal and expert costs.

There are no specific procedures relating to insurance-backed claims.

Type B-I

Roy Pembroke

The *Underpinning* case

1 The Plaintiff
The plaintiff was a private homeowner.

2 The Works
The structure of this bungalow was traditional: brick outer structural wall, cavity lightweight-block inner structural wall with internal plaster.

Its market value was some GBP 120,000.

The works were completed in February 1990.

3 Procurement Method
The property, speculatively built by a contractor member of the NHBC, was put up for sale on the open market and in February 1990 sold to the plaintiff under a standard sales contract.

4 Insurance
As the contractor was a member of the NHBC, the property was covered by the standard NHBC ten-year property damage insurance.

5 Damage or Loss
Some three years after completion, the owner, who lived in the house, discovered cracks in the masonry and the foundations.

The damage was allegedly caused by an error in the design of the foundations and faulty workmanship.

The owner claimed compensation for the cost of repairing the foundation and structure of the house, GBP 61,024, i.e. more than half the value of the property.

10 Process after Discovery
The owner made claims against his building damage insurer (NHBC, above), the contractor (and the surveyor), and also informed his household insurer.

It is not known whether the contractor or the surveyor informed their liability insurers, but it may be presumed that they did.

11 Technical Reports
The NHBC commissioned and paid for a technical report.

In the early stages of claims investigations, the NHBC and household insurers usually operate independently, although they may come together later, to negotiate a settlement under which both insurers contribute to the cost of remedial works.

At least one of the defendants also commissioned and paid for a report from an independent civil engineer.

20 Resolution
High Court proceedings were commenced but resolved by negotiation and agreement.

21 Costs
Costs would normally be paid by the losing side if a claim was resolved by litigation or arbitration.

As part of the resolution of this case, costs were shared (as per below) between the NHBC and the company that had carried out underpinning some time earlier, and whose work was found to have been faulty.

22 Information
Information was researched by lawyers and experts. It is unlikely that a claimant under the NHBC Scheme would carry out detailed research.

23 Time
The writ was served on the NHBC on November 6, 1996 and the settlement agreement was reached on May 28 1998.

24 Role of Experts
The parties used their own expert witnesses, as is the usual procedure.

At present, courts do no appoint their own experts in building litigation although it is likely under current proposals that this procedure will be adopted in the future.

Arbitrators are more likely to appoint their own experts as assessors.

25 Outcome
The NHBC and the underpinning contractor agreed to pay a total of GBP 40,000 to the homeowners; the NHBC paid 60 percent.

30 Feedback

Details of court proceedings are available to the public and the more important decisions are reported in specialist journals.

Arbitration proceedings, alternative dispute resolution procedures, including negotiated settlements, are private and no details are available unless the parties agree otherwise.

Type B-II

Deborah Brown and Patrick Perry

The *Plant System* case

1 The Plaintiff

The plaintiff was a large commercial developer with considerable experience of commissioning construction works. It was implicitly an expert in construction and had in-house construction lawyers acting for it in the dispute.

2 The Works

The developer, who owned the leasehold interest in the property, commissioned the main contractor to demolish existing buildings on the site and build a new steel-framed office building intended to be leased to business clients. The Building Services Engineer was engaged to design and coordinate the development of the building services elements of the works.

The main contractor sub-contracted the installation of the System which formed the subject matter of the dispute to the specialist sub-contractor.

3 Procurement Method

The main contractor was appointed under traditional procurement methods and engaged under an amended version of the *JCT Standard Form Building Contract, 1980 edition*. The developer, perhaps more unusually, directly invited tenders from proposed specialist sub-contractors for the mechanical and electrical works and gave an appropriate instruction to the main contractor to engage the chosen sub-contractor (although the sub-contractor was still to be a domestic rather than a nominated sub-contractor). The sub-contractor was to be engaged under the standard form *Domestic Sub-Contract DOM/1 1980 edition* with amendments, but the precise terms, and specifically terms relating to design, were arguably never finalised between the main contractor and the specialist sub-contractor.

4 Insurance

The Building Services Engineer had professional indemnity insurance, and the insurers exercised their right of subrogation to defend the claim (below). The specialist sub-contractor was also insured and was represented by the insurer's solicitors. The main contractor had insurance cover, but the claim made against it fell within the policy excess.

5 Damage or Loss

The dispute concerned part of the System installed in the building by the sub-contractor. Although used successfully elsewhere, the System was unconventional for this type of property and proved to contain incompatible parts.

Some time after practical completion, it became apparent that the System was unreliable and failed to work adequately. The specialist sub-contractor attempted to remedy the defects over a considerable period of time, but was unsuccessful, and the defects were eventually remedied by the developer, by way of wholesale replacement of the System, rather than replacement of those components which had malfunctioned because of their incompatibility with the remainder of the System.

The developer did not claim an amount in respect of consequential loss. However, it argued that the higher costs resulting from its decision to replace the whole System – rather than repair or change major components – were incurred to ensure the System worked satisfactorily, so as to avoid the potential consequential loss of business rent.

The experts generally considered the use of a particular component in conjunction with the other constituent parts of the System to have caused the damage. This underlying defect was identified only after several years had elapsed from the date of practical completion.

No individual defendant could be isolated as having caused the defect, which was likely to have resulted from a combination of poor design and selection of components by the Building Services Engineer and the specialist sub-contractor. The sub-contractor was also alleged to have contributed to the loss through incompetent workmanship. The main contractor was involved as it was contractually responsible for any default of its sub-contractor.

The damage was discovered by the developer when it commissioned the System following practical completion. The System had failed to work satisfactorily since it had been installed, and the passage of time, together with inadequate maintenance and inappropriate replacement of parts, had exacerbated the problem.

10 Process after Discovery

The developer initiated the claim following discovery of the damage.

The main contractor, specialist sub-contractor, and Building Services Engineer were all sued as jointly and severally liable for defects appearing in the System.

The developer pursued the claim on its own behalf without involving its insurer.

11 Technical Reports

The developer commissioned a technical report, which was produced by an expert at the developer's expense.

Both the main contractor and specialist sub-contractor (acting through their insurers) commissioned separate technical reports from independent experts. The respective parties paid the costs of their own experts' reports.

20 Resolution

The dispute was resolved by a one-day mediation by a mutually agreed independent and impartial mediator assisted by a technical assessor. All parties were legally represented and the mediation was attended by representatives for each of the parties, including a person from each party with authority to settle the claim.

The mediation involved a combination of technical arguments, legal discussion and direct negotiations. These were held both in open forum and in private session between each party and the mediator. The mediator and his assistant facilitated negotiations and discussion but did not give any opinion on the merits of each party's claim or defence.

21 Costs

The costs of the mediators and mediation rooms were shared equally between the parties. Each party bore its own costs of the mediation.

22 Information

The mediation occurred following service of the developer's Statement of Claim and after the three defendants had served their respective Defences. One defendant had also made a payment into court.

Perhaps unusually, there was no discovery of each party's documents prior to the commencement of the mediation process. However, each party had access to its own records on the development which would largely have included those documents which would have been disclosed had discovery occurred before the mediation. The lack of discovery did not therefore prejudice the mediation.

The developer had disclosed its technical report as part of its claim, which assisted in defining and understanding the technical issues at the centre of the case, but the other parties claimed the reports they had respectively commissioned were privileged.

It was assumed that each party had undertaken its own legal research into the strengths and weaknesses of its case.

23 Time

Following receipt of the developer's Statement of Claim, the main contractor and sub-contractor took approximately three months to serve their defences. The mediation took place after a further five months, the delay being principally due to difficulties in arranging a mutually convenient date for all the parties to attend the mediation. The mediation lasted around sixteen hours.

24 Role of Experts

The developer, main contractor and specialist sub-contractor had all instructed their own experts prior to and for the purposes of assisting at the mediation.

The parties agreed that the mediator should be assisted by a technical expert. However, his role was to assist in questioning the various parties and their experts and participating in the technical discussion to facilitate settlement negotiations, rather than to give any personal opinion on the subject matter of the dispute.

25 Outcome

The developer's claim was settled at the mediation for approximately half of the sum claimed, with agreed contributions by each of the three defendants.

No information is available as to any insurance settlement made to any of the defendants.

30 Feedback

The parties agreed that the mediation should be confidential. No details are therefore available to the general public, professional trade organisations, regulatory bodies, or the insurance industry.

The fact of settlement is evidenced by a Court Order, which is available to the public at large. The terms of the settlement reached, contained in a schedule to the Order, remain confidential.

France

Type A

J. Desmadryl

Introduction

1 Characteristics of damage observed

The superficial cracks in the external tile cladding are detrimental to the appearance of the building; the cracks, however, endanger neither its firmness nor its other features.

French case law holds quite consistently that this kind of damage – purely aesthetic and which does not imperil the building's firmness – is not within the province of producers' liability (Art 1792 of the Civil Code, often called 'Decennial Liability').

The possible liability of the parties engaged in the production of this building therefore can only be sought in the general law – 'He who, by his fault, causes a loss to another owes the other reparation' (unofficial translation).

2 Date of appearance of the damage

In order to treat the issue properly, we should add, further to the hypotheses advanced in the example, the hypothesis that the damage became visible after the Bank's acquisition of the property (and before its being offered for rent).

On the hypothesis that the damage had become visible
- Before the reception of the building by the developer: this party would only have had to make the necessary reservations or request remedial works;

Case Studies in Post-Construction Liability and Insurance, edited by Anthony Lavers. Published in 1999 by E & FN Spon, 11 New Fetter Lane, London EC4P 4EE, UK. ISBN: 0 419 24570 7

- Before the Bank's acquisition of the property: the Bank could have refused to buy, or offered a lower price, in view of the loss of value of the premises.

THE PURCHASER'S OPPORTUNITY FOR RECOVERY, THROUGH LITIGATION OR OTHERWISE, FROM THE VENDOR

The Bank's best prospect of recovering would be by calling on the vendor – in this particular case the same party as the developer –, who is also the only producer with whom the Bank has a contractual relationship. (Within the framework of the Decennial Liability, the liability of the producers – the architect, the contractor, the developer and others – is strictly transferred to the buyer, but this is not the case in general law.)

In France, in matters related to construction, the custom is, first, to lodge a complaint with the courts, even if only as a safekeeping measure pending a later settlement, whether by agreement or through litigation.

THE PURCHASER'S OPPORTUNITY FOR RECOVERY, THROUGH LITIGATION OR OTHERWISE, FROM THE VENDOR'S CONTRACTOR OR ARCHITECT

The vendor, to facilitate his defence, would surely call in the two other parties – the contractor and the architect.

Also, the Bank could seek recourse directly against them; it should, if so, prove that all, or one of them, had committed an error which had caused it a loss (in Decennial Liability trials, a plaintiff often takes action against a great number of producers, since they all may be liable to some extent, and leaves the court to allocate each one's amount of fault and hence the quota of his or her contribution to compensation).

In the case under scrutiny, it is worth noting that since the architect had, in fact, warned against the use of a certain cladding imposed by the developer, this could mitigate the architect's liability, though some experts hold that he ought to have refused to work with this developer. On the other hand, the tribunal would probably find the developer partly liable because of the doctrine of interference: the developer ought not to have imposed any particular material. The negative advice from the research centre would probably have no impact on the architect's liability, even more so as the advice appears to have remained confidential.

THE INVOLVEMENT (IF ANY) OF INSURERS IN THE PURCHASER'S CLAIM FOR RECOVERY, INCLUDING THE METHOD OF RESOLVING THE MATTER

By definition, any risk, except those created by illicit actions, is insurable, which means that for the case at issue one should first verify if this or that party is covered by an insurance which could be called upon.

In practice, in France a developer – apart from cases involving the decennial or a similar liability cover – does not normally carry insurance against risks flowing from a construction endeavour; there are insurers who offer this sort of cover, but it is not much used.

A contractor must carry the mandatory Decennial Liability insurance for the post-construction period, and nearly always insures against risks accruing during the construction period; it may even carry insurance against post-reception risks other than those of a 'decennial character' in order to have cover for actionable damage due to its faults. Likewise, an architect must, by law, carry insurance against all the consequences of its actions and errors.

ANY MECHANISMS EXISTING FOR UTILISING THE FEEDBACK OR EXPERIENCE FROM THE CASE TO A BODY OR BODIES WHICH MAY MAKE USE OF IT

The SYCODÉS data bank would be in a position to disseminate information on the case under scrutiny, particularly if the decennial liability were called upon by one of the parties involved.

In addition, insurance companies disseminate information on decisions from the courts.

Type B

F. Ausseur

Translator's note: The French Federation of Insurers – *Fédération Fran-çaise des Sociétés d'Assurance (FFSA)* – has observed that the fraction of cases which are not settled within the time limit prescribed by law (ninety days) is low (two or three per cent).

The two cases chosen here come from the small category of cases where the first settlement took more than ninety days.

In the first case, the damage insurer maintained that the damage was due to an external factor and not 'of a decennial nature', and that hence the damage insurance should not be triggered (as will be seen, the court found otherwise).

In the second case, the client had no damage insurance; damage insurance is not mandatory for the public sector, which is, in France, its own insurer.

Type B–I

The *Pavilion* case

1 The Plaintiff
The plaintiff was a homeowner.

2 The Works
The sales contract for this conventional one-family bungalow, costing some FRF 700,000, was signed in March 1983 and the hand-over took place in October the same year.

3 Procurement Method
The developer sold the bungalow to the plaintiff before completion.

4 Insurance
The mandatory *dommage-ouvrage* (D-O) insurance was taken out by the plaintiff and underwritten by Nord Stern.

The contractor carried *responsabilité décennale* insurance with SMAbtp, the Designer with GAN and the developer with Nord Stern.

5 Damage or Loss

The load-bearing walls and foundations cracked, and water infiltrated. The plaintiff discovered this in September 1989, i.e. after hand-over and during the decennial warranty period.

The damage had several causes: the foundations were not deep enough, the soil was of an inferior quality, and drought.

10 Process after Discovery

On September 25, 1989 the plaintiff filed a claim with the *Tribunal de grande instance* (TGI) of Versailles, against the contractor and its insurer SMAbtp, as well as against the developer and its insurer Nord Stern, for making good the works and monetary compensation for lack of enjoyment because of the nuisance.

The plaintiff also informed his D-O insurer of the damage, who held that the cause – drought – was 'external', and hence refused to cover the incident. The TGI called the D-O insurer.

11 Technical Reports

The plaintiff was assisted by his own technical expert, within the framework of the legal process.

The tribunal nominated an *expert* by court order dated September 28; this was paid for by an advance payment from the plaintiff.

Also the contractor's and the developer's insurers commissioned and paid for technical experts to follow up on the report produced by the court's expert.

20 Resolution

A settlement by negotiation was attempted after the court order of September 28, 1989 but without success.

The litigation went as follows:
- September 25, 1989: filing of a complaint (at the TGI of Versailles).
- September 28, 1989: The TGI ordered an *expert* to be nominated for the court.
- June 26, 1990: The TGI ordered Nord Stern and the developer *in solidum* to pay FRF 712,000 and FRF 84,443 for work, FRF 35,500 for expert costs and FRF 100,000 to be settled at the verdict.
- July 12, 1990: The TGI noted that Nord Stern had agreed to pay for remedial works.
- September 20, 1995: Verdict decreeing that the producers and their liability insurers (SMAbtp and GAN) pay Nord Stern, in subrogation for

the plaintiff, FRF 1,139,637 for remedial works, FRF 60,000 in interest and FRF 10,000 in legal, preparatory and similar expenses.

21 Costs
The costs were FRF 1,140,000 for remedial works, 10,000 in legal costs and 60,000 in interest.

22 Information
All parties had access to the process and the experts' reports.

23 Time
The plaintiff received payment for remedial works after nine months (September 1989–June 1990). The liability insurer's payment to the D-O insurer took place six years after the appearance of the damage.

24 Role of experts
See above.

25 Outcome
The TGI first ordered the D-O insurer to pay for remedial works, then ordered the liability insurers (SMAbtp, GAN) to pay the D-O insurer. The allocation of liability was 50 per cent to the contractor and its insurer (SMAbtp) and 50 per cent to the Designer and its insurer (GAN).

30 Feedback
The case is not available to the public. However, this case, together with others pertinent to foundations for bungalows, provided data for a specialist survey, by the *Centre scientifique et technique du bâtiment* (CSTB) and *Agence Qualité Construction*, aiming at changing the building regulations concerning foundations for one-family houses.

Type B–II

The *Lycée* case

1 The Plaintiff
The plaintiff was the legal representative of the *lycée* (secondary school), acting for the state (the Ministry of National Education).

2 The Works

Two-storey buildings were built as an extension to existing teaching premises.

The works started in April 1990, with hand-over on September 7 the same year.

3 Procurement Method

Tendering and procurement were according to the rules used by the state. The main contractor engaged a specialist sub-contractor for roofing works.

4 Insurance

The main contractor carried liability (*responsabilité décennale*) insurance with La Concorde and the roofing sub-contractor with SMAbtp.

The State is exempt from the obligation to take out *dommage-ouvrage* insurance, so no damage insurance was involved.

5 Damage or Loss

The parts of the works affected were the new building's flat roof, which leaked. The flaw sprung from a change in the roofing technology introduced by the main contractor without the sub-contractor being informed.

The damage was discovered by the staff of the school in June 1993, three years after hand-over.

There were at least two errors: dubious co-ordination between main contractor and sub-contractor; the subcontractor did not of his own accord modify his technique.

10 Process after Discovery

The plaintiff made a claim against the main contractor and its liability insurer (La Concorde). Also the other liability insurer (SMAbtp) was informed.

In May 1993, the sub-contractor notified his insurer (SMAbtp) of the damage, and SMAbtp commissioned an expert. This expert wrote to the expert commissioned by the main contractor's insurer (La Concorde) with an estimate of the cost for remedial works; the expert also stated that it was his opinion that the main contractor and the sub-contractor shared liability for the damage (see below).

The main contractor tendered a price for remedial work to the interior.

In October 1994, SMAbtp authorised the sub-contractor to start remedial work, but stated, in January 1995, that SMAbtp were opposed to the allocation of liability between main contractor and sub-contractor as proposed by La Concorde (90 per cent to be paid by the sub-contractor) and would accept no more than 80 per cent; SMAbtp also queried the main contractor's price for remedial works.

11 Technical Reports
The plaintiff had no report made.

Each defendant commissioned an expert to elucidate the issue. The two insurers – La Concorde and SMAbtp – requested and paid for the two reports.

20 Resolution
All issues were settled amicably, but as the experts did not agree on the cause of the damage, the payments agreed upon by the settlement were difficult to allocate.

21 Costs
See above.

22 Information
All parties had access to the work of the insurance companies' experts, and were given the opportunity to voice their opinions.

23 Time
Settling the claim took nearly two years (from the declaration of the claim and the agreement on the allocation).

24 Role of Experts
The experts employed by the defendants as per above were so-called *experts amiables*.

25 Outcome
A settlement was arranged by negotiation between insurers and producers, with 80 per cent to be paid by the sub-contractor and the rest by the main contractor.

30 Feedback
This was a particular case, which did not lead to general information to insurers or to producers (cf. above SYCODÉS).

Germany

Knut Richter and Regina Tuscher

Type A

THE PURCHASER'S OPPORTUNITY FOR RECOVERY THROUGH LITIGATION OR OTHERWISE, FROM THE VENDOR

It is an important aspect of the given facts that the building is sold to an investment bank. It has to be pointed out that in German law an isolated sale of a building usually does not exist: the subject matter of those contracts always focuses on the real estate. In this respect only the property right refers to, and is regarded as inseparable from, its essential elements – such as a building. The legal system only provides some exclusions, e.g. the *Erbbaurecht*, which might be translated as 'building lease' or 'heritable right to a building.' Accordingly, it is assumed that the vendor sold the real estate and not the building alone. Beyond this, it might be of interest to consider the fact that in the German legal system, property in real estate is transferred by the act of registration in a specific real-estate register, administered in court. In view of the fact that they have already negotiated a contract, the parties may believe that the transfer of ownership takes place at an earlier stage. This point in time is important concerning the field of responsibility.

In Germany, it is currently the case that the parties set out the conditions concerning liability and warranty periods, within the scope permitted by law, with the agreement being part of the contract as executed. These rights and obligations are regarded as non-mandatory provisions of law. Within the real estate market, it is usual to exclude the statutory warranty.

Case Studies in Post-Construction Liability and Insurance, edited by Anthony Lavers. Published in 1999 by E & FN Spon, 11 New Fetter Lane, London EC4P 4EE, UK. ISBN: 0 419 24570 7

The following consideration supposes that the statutory provisions are applied.

Claims against a real-estate vendor are to be found in the general legal provisions of purchase *Bürgerliches Gesetzbuch* (BGB), §§459ff. These provisions presuppose that a fault concerning the building is substantial. Furthermore, the fault has to cause a diminution in value of the building, or restrict its fitness for its purpose. If so, the purchaser may, within certain periods, insist on rescission or the later reduction of the purchase price.

German courts have developed a definition of fault: the state of a physical object must
(a) differ from what was meant by the contracting parties or be regarded as differing in the sense of general or generally accepted standards; or
(b) significantly restrict the fitness for its purpose or the value of the object.

The facts of the case considered do not provide sufficient details to deal with these criteria. Superficial hairline cracks do not represent a limitation of the building's fitness for purpose as such, because they do not affect its stability or other performance.

(a) However, the aesthetic impairment of the façade by hairline cracks might be considered a 'fault'. Evidently, the impairment of the façade is serious enough to cause a negative impression of the building owned by the Bank. It may also be argued that there is a fault because the state of the façade obviously differs from the one expected by the parties.

It was only because the developer's senior staff held strong views on the desired aesthetics that the untried product was used. It may therefore be concluded that the appearance of the building was important to the developer. In this situation, one may assume that the parties considered the aesthetic appearance to be a term or object of the contract; but, as it later turned out, the state of the façade differed from the state meant at the time of the conclusion of contract.

Another line of reasoning actually leads to the same result: as the parties did not mention any specific detail concerning methods and materials, it may be assumed that the Bank concluded the contract supposing that construction had been performed with sufficiently tested methods and acceptable materials. This would, again, mean a divergence between the actual and the supposed state.

(b) The fault must diminish the value of the building or reduce its fitness for purpose.

A reduction in the value of the building may be assumed, but the facts provided do not allow a complete decision on that point. There are no hard and fast principles according to which a solution may be reached; the courts individually decide questions concerning reduction of value.

Furthermore, in respect of a warranty claim, it must be considered whether the purchaser knew about the fault at the time of purchase (BGB §460). Where the purchaser knows about faults in the subject matter at the time of purchase, the vendor carries no liability in respect of these. If the purchaser does not know about them because of his own gross negligence, the vendor is only liable to the purchaser in the event of malicious non-disclosure.

According to the facts given, representatives of the Bank did not visit the building; one therefore concludes that the purchaser did not take any notice of the façade. The facts given do not tell whether the Bank did not know about the flaws because it did not inspect the building, or because the flaws appeared only after the purchase.

If the purchaser did not know about the flaws because of its own gross negligence, the vendor would only be liable in case of malicious non-disclosure.

It may be concluded that the Bank had no knowledge of the hairline cracks because of its negligence: it is accepted practice to inspect a building before purchase. Because it is not clear from the facts given whether the fault could have been noticed at the time of purchase, the following consideration is based on the assumption that the purchaser could have done so.

The second legal precondition of this stipulation requires malicious nondisclosure by the vendor.

According to a rule referring to a decision by the *Bundesgerichtshof* (the Supreme Court), the vendor is liable if he has positive knowledge of the fault or ought to have been aware of it, and does not inform the purchaser. The facts given reveal that the study about the product was circulated after the time of the purchase among experts, and that the vendor acted in good faith.

In view of this, the purchaser would have the right to demand

- A rescission of the transaction,
- A reduction of the purchase price, or
- Compensation for non-fulfilment of the contract.

As to the computation of the price reduction, practice offers different procedures. One of the opinions is based on the amount of necessary repair.

Given that, at the time of purchase, the real estate (including the building) was deficient with respect to a quality which was warranted, compensation may be claimed.

In German law, aesthetic appearance as such may not be considered a quality which is warranted. Even if a certain appearance could be considered to have been warranted, the facts given do not show that the parties agreed on any specific appearance, either implied or explicitly.

The general rules created by case law referring to a 'breach of a secondary contractual obligation' do not lead to the vendor's liability. The facts given furthermore leave no room for considering a tortious claim against the vendor.

THE PURCHASER'S OPPORTUNITY FOR RECOVERY THROUGH LITIGATION OR OTHERWISE, FROM THE VENDOR'S CONTRACTOR OR ARCHITECT

Usually, real-estate contracts provide an assignment of warranty claims that transfers these rights from the vendor to the purchaser. In such cases, the purchaser could turn directly against the contractor or the architect. The case considered does not suggest any possibility of this alternative.

The fact that no contractual link exists between the purchaser and the vendor's contractor or architect excludes any contractual claim, such as warranty claims.

Again, there is no hint in the facts given of any possible basis for liability in tort. Thus, no ground for a direct legal claim by the purchaser against the vendor's contractor is to be found.

THE INVOLVEMENT (IF ANY) OF INSURERS IN THE PURCHASER'S CLAIM FOR RECOVERY, INCLUDING THE METHOD OF RESOLVING THE MATTER

Although warranty claims are considered part of the performance of a contract, they are not insurable in the sense of protection against sudden and unforeseeable damage. In Germany, there is no normal insurance in this field, and therefore no insurance product exists on the market that would protect a purchaser's claims in respect of the alleged breach of contract.

THE PROCEDURES BY WHICH THE ABOVE ARE RESOLVED, INCLUDING EXTRA-LEGAL, LEGAL AND INSURANCE PROCEDURES

(a) Extra-legal procedures

If the parties have included an arbitration clause in their contract, they may resolve their conflict by applying the procedure thus agreed upon. Usually, this would still offer the possibility of a civil action as a second stage.

To avoid long and expensive court processes, a solution might lay in negotiating a compromise by concluding an extra-judicial settlement. In this case, a court could be involved in case of problems with the terms of execution.

(b) Legal procedures

The usual way of dealing with construction conflicts would be by filing for a civil action.

(c) Insurance procedures.

In the absence of a specific insurance of this kind, there is no procedure to be mentioned.

ANY MECHANISMS EXISTING FOR UTILISING THE FEEDBACK OR EXPERIENCE FROM THE CASE TO A BODY OR BODIES WHICH MAY MAKE USE OF IT

There is no formal mechanism to describe. Spectacular cases would be published in professional journals for experts in construction or law. In addition to that, reports might be found in publications by the *Bundesamt* or the *Landesämter für Materialprüfung*, if the product had been subject to approval procedures.

Type B–I

The *Basement Garages* Case (fictitious name), relevant to the jurisdiction of Lower Saxony.

1 The Plaintiff

The plaintiff, a developer with experience in the field of residential projects, planned to erect a nine-unit building on land he owned, intending to sell the apartments and basement garages unit by unit, before or during construction.

2 The Works

The developer commissioned an architect to design the project, estimate costs, obtain all permits needed from the authorities, choose contractors and supervise construction.

Having studied the plans and specifications drawn up by the architect, the authorities issued the necessary permits, and the developer started selling the units, on the basis of the plans and specifications. The developer fixed the price of each unit on the basis of the architect's estimates.

The developer promised to transfer the properties to the purchasers within a certain period after the formal hand-over of the works.

3 Procurement Method

Foundation works were performed by a specialist contractor.

4 Insurance

The architect carried professional liability insurance.

5 Damage or Loss

During the preparatory excavation, collapse of a wall and cracks in a building on a neighbouring site occurred.

The error was in the design.

To prevent further damage to the neighbour's property, the architect modified the plans to accommodate supports during excavation, thus increasing the distance to the neighbour's land. These safety measures led to a reduction in the size of the garages which consequently became smaller than indicated on the plans used for the sales (above).

The works were completed according to the modified plans.

The garages being smaller than promised in the sales contracts, the purchasers of the units successfully claimed out-of-court compensation from the developer. The developer also suffered a loss by having to pay for the unforeseen supporting appliance during excavation.

The developer learned about the architect's modification of the plans only after the garages had been completed to the reduced dimensions.

10 Process after Discovery
The neighbour claimed compensation for damage to his property against the developer; this claim is not dealt with here.

As mentioned, the purchasers of the units claimed out-of-court compensation from the developer. The developer then claimed financial compensation from the architect for this, alleging that the architect had modified the plans without proper authority (the size of garages) and without informing the developer.

The developer also claimed for the cost of the safety measures, alleging that it was the architect's duty to investigate soil conditions properly before planning, and to include the safety measures in the estimates, which would have made it possible to consider them when the price for the units was calculated.

11 Technical Reports
The developer filed a motion for a *Beweissicherungsverfahren*, a special procedure that implies suspension of the limitation of actions, because he wished to ascertain whether the supports had been necessary and whether it was unavoidable to reduce the size of the basement garages; the court accepted the motion and two days later issued an order to this effect, and requested that the expert proposed by the plaintiff – a qualified engineer chosen from an official list of experts listed by the Chamber of Engineers – be nominated.

The expert opinion obtained stated that the soil at that part of the site contained an untypical and irregular layer of clay, and that the supports had had to be erected to avoid further damage to the neighbour's property.

The developer in his role of applicant had to pay in advance for the expert opinion according to the B*eweissicherungsverfahren* procedure.

The defendant did not commission a report, and in the following litigation procedure referred to the expert opinion put forward in the *Beweissicherungsverfahren*, which, as mentioned, supported the architect's choice of technology.

Furthermore, the tribunal (below) requested additional expert opinion as to whether the architect should have been able to recognise the soil conditions, and whether it had been within the duties of the appointment to investigate these.

20 Resolution

The conflict between the developer and the purchasers (above) was settled by negotiation, with a reduction in the price of the units.

The litigation between the developer and the architect went to court. The tribunal referred both to the result of the expert opinion stated in the *Beweissicherungsverfahren* and to the further report commissioned (above). The latter concluded that for geotechnical reasons it would have been impossible to build the garages as originally planned.

The tribunal found that it could not be concluded that the architect should have ordered an investigation of the soil conditions, even though the geological situation was not typical for the area; no such contractual duty had been concluded with the developer.

The tribunal found that the architect was in breach of duty by not notifying the developer of the modification of the size of garages, but that the architect was not liable for the compensation claimed by the developer.

21 Costs

All costs, including the cost for the defence, were to be paid by the plaintiff, who lost the case. The cost of the *Beweissicherungsverfahren* had already been paid by the developer.

For financial reasons, the plaintiff did not pursue an appeal.

The case became an expensive process because of the costs for expert opinions (approximately 25 per cent of the amount in dispute).

22 Information

All parties to the conflict had access to all studies, research and data concerning the issue needed to plead their case (the purchasers had no access to the expert opinions).

23 Time

From discovery of defect until the plaintiff's motion for the *Beweissicherungsverfahren* two weeks elapsed. The expert opinion was produced within another four weeks.

The litigation took place one year after discovery (in the meantime the negotiated agreement between the developer and the purchasers was concluded).

24 Role of Experts

Experts played essential roles in this technically complex case; this may be said to lead to the conclusion that fact-finding with experts' help becomes the real instrument for the court's ruling.

Also, the plaintiff's right to propose an expert within the *Beweissicherungsverfahren* procedure, as well as the tendency of courts to nominate the expert thus proposed, could be considered to create a precedent.

25 Outcome

As mentioned above, the plaintiff completely lost the case.

The architect's insurer had been notified at the beginning of the conflict. Since the architect was not held liable, the professional liability insurer was not involved in any terms of payment.

30 Feedback

Significant selected cases are usually published anonymously in official court anthologies or in law journals.

Type B–II

The *Electricity-Smog* Case

Relevant to the jurisdiction of Freistaat Sachsen (Saxony), one of the 'new' German *Bundesländer* where a significant pent-up demand for private homes remains. For this reason, 'new' *Bundesländer* run financial support programmes.

1 The Plaintiff

The plaintiff intended to erect a one-family bungalow and commissioned an architect to design the house and to ensure that the client received a subsidy from one of the financial support programmes of the *Land* of Saxony.

2 The Works

The architect designed the building. The architect had come from an 'old' *Bundesland* to practice in Saxony and was not well informed about details of the several home-building support programmes available.

Meanwhile, the plaintiff purchased land from the local authority. This land had not yet been surveyed (in order to further the construction of private homes, the local authority sold tracts of land, though precise property boundaries were to be identified only after a final land survey. In the 'new' *Bundesländer*, there are still many cases with unclear identification of ownership of real estate).

Within the area, there was a high-tension power cable. Several bodies claimed to have legal rights to the cable.

3 Procurement Method
Not relevant.

4 Insurance
The architect carried professional liability insurance.

5 Damage or Loss
The final land survey showed that parts of the land sold to the plaintiff were under the power cable. When the plaintiff saw the land, he became aware of its unattractive location. The architect had not visited the site after the final survey.

When the architect had read the conditions for subsidy, he understood that the deadline for applying had elapsed. Because of this, the client could no longer benefit from any such programme.

The underlying errors were the following:
- Concerning the claim against the local authority: location of the land allocated.
- Concerning the claim against the architect: missed deadline for applying for a subsidy.

10 Process after Discovery
In the first instance, the plaintiff intended to claim that the power cable be removed. Before he could file for this, he was informed that, because the issue of ownership to the electric cable was not yet settled, it was unclear who would be in the role of defendant.

The plaintiff then demanded rescission of the contract for his purchase of the land, and compensation for the fee he had already paid to the architect. Furthermore, he claimed compensation from the architect for the unachieved subsidy.

The plaintiff commenced litigation against the local authority, claiming compensation for his payment of the architect's fee, which had been reduced to the amount which he could have obtained through a subsidy. The plaintiff held that the land would not be suitable for building a home because of the proximity to the high-tension cable, which, he was convinced, would cause health problems.

The plaintiff also sued the architect for the moneys he could have obtained by applying for a subsidy in time.

11 Technical Reports

The plaintiff had to obtain a private expert opinion in order to answer the question whether living close to a power cable would have a negative influence on health through the phenomenon known as 'electric smog.' He failed to find an expert who would report on this because of the absence of scientific proof of a causal link between health disorders and living close to a high-tension power cable. Also the tribunal tried, and likewise failed, to clarify the question of a possible health risk created by 'electric smog.'

An expert was commissioned by the plaintiff to elucidate the question as to whether the building planned could have been erected on the land allocated. It was found that in the absence of sufficient legal stipulations for construction in Saxony, there was no administrative or regulatory reason to oppose such a project. Likewise, the internal stipulations of the power supplier raised no objection to the housing project.

20 Resolution

Prior to the litigation concerning the reimbursement of the architect's fees, negotiation between the plaintiff and the local authority had led to a solution of the first issue, namely that the authority took back the land and reimbursed the price paid. This agreement could be understood as recognition of the claim that the location of the land was unsound.

21 Costs

Further to the legal costs, there were notarial charges for the negotiated rescission.

22 Information

All parties had access to the same information.

23 Time

The period between the purchase of land and the survey was around eight months. After a further month, the plaintiff identified the final property boundary, which showed that his building would be located very close to the power cable. The settlement with the local authority followed approximately two weeks after the plaintiff had become aware of this circumstance.

The litigation procedures against the architect and against the local authority were initiated eleven months after the purchase.

The process concerning the local authority was closed after fourteen months, mainly because, as mentioned, no expert could be found.

The case of the plaintiff against the architect took two months.

24 Role of Experts

Only private expert opinions were asked for, by the plaintiff, concerning whether the erection of the building was in accordance with official planning regulations.

25 Outcome

As mentioned, the plaintiff and the local authority reached a settlement as to rescission.

The action concerning compensation for part of the architect's fee was dismissed by the court.

In default of a scientific opinion about 'electricity-smog' and human health, the tribunal saw no justification for a claim; however, it did not rule out future action.

The plaintiff succeeded in his action against the architect, who had tried to exculpate himself on the ground that he usually worked in another *Bundesland*, and who advanced, without success, the argument that administrative responsibilities in the new *Bundesland* of Saxony were so badly established that he could not avoid failure.

The architect's liability insurer paid the cost of his legal defence, the cost of litigation incurred by the plaintiff and the court fees. In addition to that, the insurer paid compensation to the plaintiff according to the contractual franchise clause. The insurer decided not to appeal.

Regarding the negotiated settlement with the local authority, the costs for notarial charges were shared equally by the parties.

In his unsuccessful claim against the local authority for compensation for the architect's fee, the plaintiff had to pay court costs and both parties' legal fees.

30 Feedback

This case is not available to the public.

Hong Kong

Type A

*Edwin H. Chan with Mohan M. Kumaraswamy, Robert J.M. Morgan
and Kumaru Yogeswaran*

ASSUMPTIONS

The following assumptions are made in accordance with the local situation:

In the conveyance, the vendor made no representation as to the quality of the cladding tiles. The conveyance was based on the *caveat emptor* (buyer beware) principle, where the purchaser had to take the property as he found it.

A warranty as to the quality and fitness of the cladding tiles was given by the supplier to the developer.

THE PURCHASER'S OPPORTUNITY FOR RECOVERY, THROUGH LITIGATION OR OTHERWISE, FROM THE VENDOR

A claim for breach of contract would be ill-founded as the sale would be based on the *caveat emptor* principle, meaning that the purchaser was presumed to have accepted all defects in the building that he could have discovered had he commissioned an experienced surveyor to do a survey. This would be normal practice for a property transaction of some value in Hong Kong. Hence, the vendor had made no misrepresentation.

Case Studies in Post-Construction Liability and Insurance, edited by Anthony Lavers. Published in 1999 by E & FN Spon, 11 New Fetter Lane, London EC4P 4EE, UK. ISBN: 0 419 24570 7

The vendor, having pre-empted the architect's professional opinion, might be worried about his liability for negligence, but the standard of care required of him would be much lower than that of a building professional. The damage would be regarded as a development risk, and would be reasonable even for a prudent developer to be ignorant of the defect inherent in the tiles. Having acted in good faith might allow him to avoid liability for negligence.

Even if the vendor were found negligent, the purchaser's loss was purely economic, and Hong Kong courts would not normally award damages for this unless a special relationship existed between vendor and purchaser.

THE PURCHASER'S OPPORTUNITY FOR RECOVERY, THROUGH LITIGATION OR OTHERWISE, FROM THE VENDOR'S CONTRACTOR

The purchaser would not be able to claim against the contractor in contract because of lack of privity, nor in tort because the contractor was not negligent in using materials specified in the contract.

THE PURCHASER'S OPPORTUNITY FOR RECOVERY, THROUGH LITIGATION OR OTHERWISE, FROM THE VENDOR'S ARCHITECT

Again, the purchaser could only claim against the architect in tort for negligence in specifying the cladding tiles. The defence of the architect would be that he had carried out his professional duty with reasonable care: He had rejected the tiles because he and other practitioners in the construction field had no knowledge of them and could not obtain reliable information about them, and the tiles were used by the developer despite the architect's explicit objection, which would constitute proof of reasonable care on his part.

Since the report from the building research institute was not available at the time, the architect was not in breach of the standard of the ordinary competent architect in not being aware of the defect in the tiles. If the developer did not follow the architect's advice, he was not required to take on the extra duty to carry out research concerning the tiles.

Even if the architect were found liable to the purchaser, the vendor would be held to indemnify the architect as merely carrying out specific instructions from the vendor, acting as a principal.

THE INVOLVEMENT (IF ANY) OF INSURERS IN THE PURCHASER'S CLAIM FOR RECOVERY, INCLUDING THE METHOD OF RESOLVING THE MATTER

There is no insurance policy covering latent defects in Hong Kong. Normally, no insurance policy would cover warranties, unless it was specifically extended for an extra premium.

Insurers in this case could include:

(a) Insurers of the contractor and insurers of the supplier for their respective contractual performance under the building contract;

(b) Insurers of the vendor and insurers of the purchaser, both covering specific perils, such as fire, explosion, etc.; and

(c) Insurers of the architect for professional indemnity.

When the architect issued the Final Certificate under the building contract, the contractor would have proof of having satisfactorily performed his contractual obligation, save for fraud. A contractor's insurance policy would not normally cover workmanship and definitely not cover fraud. The contractor's insurers would not involve themselves in the purchaser's claim in this instance.

The supplier's insurers would have a vested interest in the claim, although the policy might not cover the warranty. Upon notification of the possible claim, the architect's insurers would also be interested, just in case their insured incurred liability.

The vendor's insurance policy would have expired upon sale of the property

Depending on the specific perils in the purchaser's policy, which might include the maintenance of the facade, the purchaser's insurers might be interested in the claim. The insurers would not be parties to the claim. However, the parties would be required to notify their insurers upon receiving any claims.

Insurers would advise their respective insured parties to negotiate and instruct solicitors, unless the insured parties object with good reason. They would first attempt to avoid any liability at all, or any liability above the excess in the policy. If they could not avoid this, they would attempt settlement by negotiation, mediation or arbitration.

THE PROCEDURES BY WHICH THE ABOVE ARE RESOLVED, INCLUDING EXTRA-LEGAL, LEGAL AND INSURANCE PROCEDURES

All the above parties claiming and being claimed against could join in as parties in the dispute. Negotiation for settlement would be carried out, in the parties' or in the insurers' offices.

If no settlement were achieved, mediation would be carried out by a mediator agreed to by the parties. There are no rules or methods to be followed, but the Mediation Rules of the Hong Kong International Arbitration Centre (HKIAC) would either serve as a reference or be adopted.

If it was still not possible to achieve a settlement, and both parties still wanted an out of court settlement, they would agree on an *ad hoc* arbitration agreement. A sole arbitrator would then normally be appointed by the parties or, failing that, an appointment would be made on the recommendation of a local insurance institute. The arbitration procedure would be more formal and would have to satisfy the Hong Kong Arbitration Ordinance. The procedure would follow the HKIAC Domestic Arbitration Rules or the HKIAC Short Form Arbitration Rules. If the arbitration award were not honoured, the winning party could seek to enforce the award in the High Court.

Due to the nature of the damage, any claim in tort would have great difficulties in succeeding. The most straightforward claim would in principle be based on the supplier's possible warranty, but this might not be available to the purchaser, since rights under it may not have been assigned. The remedy available might also be restricted by the terms of the warranty.

Hence, the dispute would most likely be settled through negotiation, with a major contribution from the supplier for its defective product, and the contractor for its workmanship in installing the tiles, and also possibly some contributions from the vendor and the architect, all for commercial reasons. According to the normal rule that 'costs follow the event', should the dispute be referred to arbitration, the losing party would have to pay for the cost of the reference and the arbitrators' fees only for the issues that it had failed to defend successfully.

ANY MECHANISMS EXISTING FOR UTILISING THE FEEDBACK OR EXPERIENCE FROM THE CASE TO A BODY OR BODIES WHICH MAY MAKE USE OF IT

Unfortunately, there is no mechanism in Hong Kong for utilising the feedback or experience from cases like this. If the case were heard in Court, it would be covered by the press or the law reports or both. Otherwise, feedback would be through gossip in the trades and professions. There is no national building research institute in Hong Kong to distribute technical information. If the implications are very serious, some of the professional institutes may send out flyers to alert their members. `

Type B–I

Robert J M Morgan, Edwin H Chan, Mohan M. Kumaraswamy and Kumaru Yogeswaran

The *Plinkton Estate* case (fictitious name)

1 The Plaintiff
The plaintiff was a developer, member of a Taiwanese owned group of companies specialising in high class residential and commercial developments both in Hong Kong and in Taiwan.

2 The Works
The project was a luxury residential estate and resort, including a hotel, restaurants and extensive leisure facilities. The land value before development was some HKD 1 billion, with an estimated value at completion of HKD 6 billion.

The works were commenced in 1984 and were completed in 1989.

3 Procurement Methods

Architecture and structural engineering services contracts were negotiated with major, well-established Hong Kong firms. The Developer specified nearly all materials.

Building works were procured by a negotiated contract with a contractor, a major Japanese corporation, selected by the Developer, with no tendering procedure. The form of contract used was prepared specifically for this project, and was based upon a number of standard forms, including the Hong Kong Government Standard Form of Building Contract. The Contractor was, among other things, required to provide marble to match selected samples from a particular quarry. Thus, although the supplier was not nominated under the contract, it was in fact nominated because one party controlled the choice of quarry.

The Developer also nominated nearly all major subcontractors. Tiling (below), however, was done not by a nominated sub-contractor but by sub-contractors chosen by the Contractor.

4 Insurance

The Contractor carried a Standard Contractor's All Risk Policy. The Designers all carried Professional Indemnity Insurance.

5 Damage or Loss

Complaints arose at an early stage but were ignored by the Contractor (see below). They mainly concerned

- The structure :
  Concrete badly out of tolerance and honeycombed.
- Internal Fittings :
  Woodwork and marble not of the quality or type specified,
  HVAC equipment not up to the standard required.
- External Finishing: Major problems with cladding tiles coming unstuck.
- With the risk of tiling collapsing onto passing pedestrians and vehicles below, this proved the most dangerous and contentious problem.

The damage was the result of poor workmanship and supervision by the Contractor. The errors were noted by almost all members of the Developer's consultants and construction supervision teams, including the Clerk of Works, the Architect and the Structural Engineer. The Architect and other of the Developer's consultants complained, and certificates and opinions were issued which noted the Contractor's failure to comply with the Contract's specifications.

10 Process after discovery

The issues outlined above initially led to the consultants' refusing to issue interim certificates – which reflect 'the value of work properly carried out' during a certain period, usually one month – and hence to deductions from payments to the Contractor; after a few months, the Developer refused further payments. The Contractor, under protest, continued to carry out work, but ongoing problems culminated in the Developer's terminating the Contractor's employment at a stage when approximately 75% of the works had been completed.

As the Contractor had disregarded the consultants' advice, no claim was made against them. It should also be noted that the consultants had had no say in the choice of contractor.

10 Process

When the magnitude of the problems was fully appreciated, the Contractor's engagement was, as mentioned, terminated and a new local contractor was invited to complete the works on a Cost-Plus Basis (since there was considerable debate as to the extent of the remedial works necessary, it was not possible to tender or negotiate for the completing works).

11 Technical Reports

Both the Contractor and the Developer obtained their own technical reports, at their own expense. Expert advice was sought from leading international experts in tiling and tile adhesion; numerous technical reports were produced.

No insurers were involved.

At one stage the experts maintained that in order to ensure safety, it would be necessary to remove and replace all tiling. This would have been a very costly and logistically almost impossible task (owing, *inter alia*, to the huge amount of waste material involved). The Developer's claim against the Contractor was, however, based on the cost of (a) the entire replacement of the tiles progressively over a ten-year period, and (b) the cost of completion by another contractor.

20 Resolution

The matter was resolved partly through arbitration and eventually through a negotiated settlement between the Developer and the Contractor.

There was no 'approved' tiling method in the Contract. The Contractor was under a general common law duty to carry out works in a reasonably workmanlike manner and so that they would be fit for their purpose. Failure to follow the tile-adhesive manufacturer's instructions, it was found, constituted a failure to carry out the works in a reasonably workmanlike manner. In the arbitration proceedings (below) the arbitrator held that the obligation on the Contractor to execute works 'regularly and diligently' were not limited to progress but also referred to the manner of working.

In the event, replacement of tiles as and when necessary became an ongoing process. This, combined with the use of strategically placed canopies to protect pedestrians and vehicles from falling tiles and debris, made it possible to occupy the buildings, thus producing an income for the Developer, albeit with extraordinarily high maintenance costs.

Both parties made claims against the other: The Contractor claimed approximately HKD 300 million, being the value alleged for completed works and the loss of profit on the uncompleted portion, while the Developer's claim amounted to some HKD 6 billion, being the cost of completion by others, loss suffered by delayed occupancy and letting of the buildings, increased maintenance costs for progressively replacing the tiling etc.

No contribution was claimed from any consultant.

As the contract contained an arbitration clause, the matter was referred to arbitration. Both parties were at that stage represented by major firms of Hong Kong solicitors. An arbitrator, a senior UK lawyer (a Queen's Counsel), was agreed upon and appointed by the parties jointly.

The Contractor, as well as being involved in the main proceedings with the Developer, became involved in a number of subsidiary arbitration proceedings with sub-contractors.

The proceedings were among the largest ever in Hong Kong, and probably in the world. At its peak, the main arbitration involved some 30 construction lawyers, specialists in various aspects of the dispute. Due to the scale and wide-ranging nature of the dispute, the main arbitration progressed through a number of interim awards dealing with specific matters and issues. At the time of settlement, two interim awards had been issued, a third was about to be issued, a fourth hearing was scheduled, and one application for leave to appeal was about to be heard in the High Court.

21 Costs

The losing party normally has to pay its own and a large proportion of the winning party's cost. In this particular matter, both parties had been successful in some issues contested in the hearings, and hence both had been awarded some of the costs incurred in arguing certain matters. The costs position, therefore, is too complex in terms of who has 'won' the arbitration and who has 'lost'.

22 Information

Generally, both parties had access to their own and the other side's documents relating to this matter. Under Hong Kong law, 'discovery' of documents is usually ordered by an arbitrator in construction cases and takes two forms:

- 'General' discovery, whereby each party is required to produce for inspection documents (other than documents covered by some form of legal privilege, such as correspondence aimed at compromising the dispute) set out in a list of documents identifying relevant documents in the possession or power of each party. A 'relevant' document for this purpose is one which supports a party's case or weakens that of his opponent, or which leads to a train of enquiry having either of these consequences;
- 'Specific' discovery, which may be ordered against a party who fails to make sufficient disclosure in a 'general' discovery. The Developer was marginally disadvantaged because some of the documents were in the hands of sub-contractors. There are significant difficulties in situations such as these in gaining access to documents held by third parties, particularly where those third parties themselves may have some interest in the outcome of the proceedings. In this situation, the High Court, in support of the arbitrator, may be asked to issue a *subpoena duces tecum* ordering a Third Party to appear before the arbitrator to give evidence and produce documents.

This procedure, however, was not used.

23 Time

As mentioned, the arbitration involved a vast number of issues and had been under way for some two and a half years at the time of settlement. The proceedings were projected by one of the parties to last for 10 years. The other party was keen to reach resolution as quickly as possible and hence wished so far as possible to expedite matters, projecting five years overall as the maximum acceptable period for reaching a final award.

24 Experts
Both parties appointed their own experts; no expert was appointed by the tribunal.

25 Outcome
The matter was, as mentioned, eventually resolved by means of a negotiated settlement, the details of which are confidential. The settlement, furthermore, involved wider commercial matters than the contract in dispute.

As mentioned, no insurer was involved.

30 Feedback
Arbitration is a strictly private process and publication of details of the arbitration, the arbitrator's awards or any settlement whereby the parties may be identified, is prohibited without the consent of both parties and the arbitrator. Accordingly, the outcome and details of the proceedings are not available to professional or trade organisations, regulatory or standard-setting bodies, the insurance industry, or indeed to anyone else.

Type B–II

The *Railway Station Planks* case

Kumaru Yogeswaran, Mohan M. Kumaraswamy, Y. Ueda, Edwin H Chan and Robert J M Morgan

1 The Plaintiff
The plaintiff was the client, a Government owned Corporation (public sector).

2 The Works
Manufacture and installation of pre-cast concrete planks with a cover to reinforcement of 15 mm with a tolerance of 5 mm either way (i.e. less than 20 mm and more than 10 mm cover).

The planks were manufactured by ST, a building materials supplier and manufacturer, at an off-site pre-cast concrete yard, and installed in a railway station complex by AB, a joint venture of two contracting companies.

The technology was not innovative but based on a Code of Practice (below).

3 Procurement Methods

- *Client – Contractor*: The Client's *Conditions of Contract for Civil Engineering and Building Construction*, March 1981, applied. Such contracts are subject to a six-year limitation period from the date of substantial completion for initiating any action for breach of contract.
- *Designer*: The terms of employment for this independent consulting practice may have been identical or similar to the standard terms of the Association of Consulting Engineers.
- The Contractor (AB) entered into a *Sales and Purchase of Goods Agreement* with a *manufacturer and supplier* (ST), for 'pre-cast concrete Grade 30/20 reinforced slabs.'
- *'Reviewing Engineer'*: Appointed by ST to monitor repair work carried out by another contractor on behalf of the Client.

4 Insurance

The Client carried comprehensive insurance cover for design and construction (a similar insurance policy, termed *Owner Controlled Insurance Policy* (OCIP) was established by the Hong Kong Government for the Airport Core Programme. OCIP covers Contractor's All Risk Insurance, Marine Cargo Insurance and Third Party Liability Insurance). The Client, who paid the premiums, informed its contractual parties of the policy's excess clauses.

The Contractor was required to take out policies covering contractor's equipment, marine craft insurance, motor vehicle and aircraft insurance, employee's compensation insurance, and professional indemnity insurance.

The Designer was required to obtain Professional Indemnity Insurance.

5 Damage or Loss

In 1987, nearly two years after completion, the following flaws in the pre-cast planks were noted by the Client: spalling, cracks parallel to and over reinforcement bars, honeycombing or other porosity of concrete over the reinforcement.

10 Process after Discovery

In January 1988, the Client confirmed, in writing, that an inspection was to be implemented, the next month, jointly with AB and ST in order to identify the flaws and underlying defects mentioned under 5, and 'whether the concrete cover was less than 10 mm [see below], and any other defects which may be agreed upon.'

In September the same year, the Client advised the Designer, AB and ST that spalling was noted also on planks not yet installed, and proceeded with further tests to ascertain whether there were any reasons other than too shallow a cover to the reinforcement bars for this damage. To test for chloride content and to check for carbonation, the Client collected samples of corroded reinforcement bars and concrete dust.

11 Technical Reports

Six planks were selected to test for the presence of chloride (which could explain the damage). In December the same year, the Client advised the other parties that the tests had not revealed the presence of chloride (that seawater may have been used in pre-cast planks had been alleged in some other projects around that time).

ST advised AB on September 24 and November 14, 1988 that the Code of Practice in force, CP 110 (Part 1, 1972, para. 3.11.2 and Table 19) stipulated a nominal cover of 30 mm (Grade 30 concrete, moderate exposure), while the cover specified in the contract was only 15 mm, and that ST believed that this was the major reason for the spalling observed.

At the same time, the Client was proceeding with arrangements to have the damage repaired by another contractor by coating all planks with less than 15 mm cover. The technical specifications for that contract were prepared by the Designer who also drew up a list of tenderers. The cost of the treatment was estimated at HKD 3.1 million.

The Client planned to award the contract by March 1989, and in February tenders were submitted. All parties were invited to participate in the assessment of tenders.

The Client also replied to AB's comments on liability for the spalling:

* *Chloride in rust and diffusion of chloride from the bars*

Analyses by Scanning Electron Microscope did not indicate a distinct profile, and the level of chloride found was compatible with ordinary concrete with no chloride contamination and no spalling. Hence, it was considered reasonable to associate the chloride with the steel, rather than with the concrete.

- *Chloride on steel*

Based on the test results, the chloride content was, as mentioned, not excessive. The conclusion was thus that the major factor in causing corrosion and spalling to occur at such an early date after manufacture was a deficiency in concrete cover.

- *Design Specification, Concrete Cover and Tolerance*

The exposure was considered 'mild', and the contractually prescribed cover of 15 mm was therefore appropriate. The structure was subject to Hong Kong Building Ordinance and Building Construction Regulations, which specify a minimum of 15 mm cover, with no mention of tolerance. Hence it was considered necessary to restore the concrete cover to a depth of at least 15 mm.

- *Client's Supervision*

The presence of the Client's representative during manufacture and installation did not relieve a contractor from his responsibility to provide the finished works in accordance with the specification.

- *Errors in Drawings*

The Client acknowledged that some drawings issued at the commencement of the works contained errors concerning cover. However, the incidence of shallow cover appeared to extend throughout the works.

- *Allocation of cost of remedial works*

The Client gave his opinion that factors concerning workmanship, material and design were at the root of the problem, and that the cost of remedial works therefore should be the subject of discussion and agreement to be reached between the Client, the Designer and AB.

In a memorandum of March 4, 1989 the Client stated that

- an estimated 25 per cent of the planks were cast with less than the required 15 mm concrete cover;
- workmanship was the main cause of the damage;
- the Designer did not agree with AB's allegation that the design was inadequate (AB had pertinently pointed out that the maximum aggregate size was 20 mm, i.e. greater than the cover prescribed);
- neither AB nor ST raised the alleged design inadequacy at the time the units were manufactured or installed.

AB, in a reply dated March 6

- contended that results of the testing should be made available before any discussion of cost distribution could take place;
- stated that ST had initially agreed, as a gesture of goodwill and without admitting liability, to offer a cash settlement;

- said that since the cost of repairs had increased, ST wished to establish the reasons for the flaws;
- confirmed that AB's position was similar to ST's, and that AB also was committed to honour 'its responsibility;'
- wished to know whether the Client intended to investigate the avenue of insurance since the Client had a comprehensive insurance cover for design and construction (4 above);
- advised that they would not accept any proportion of the cost prior to a full investigation of the test results and before the reasons for damage to the planks had been established.

On March 13, ST advised AB that they were of the opinion that the problems would not have occurred had sufficient cover been specified. ST also said that as the estimated costs had now increased, they no longer agreed to a cost sharing agreement.

The Client subsequently stated that

- irrespective of whether the cover was inadequate or not, AB, and in its turn ST, were obliged to produce planks with the specified cover;
- since the damage was related to workmanship, no cover from insurance would be available;
- remedial works were required to rectify the defective works only.

The Client also produced a table totalling 1163 defective planks with identified problems; another 903 planks were regarded as potentially defective, giving a total of 2066 planks in need of repair.

On November 10, the Client presented to AB, ST and the Designer the following estimate for the repair (in HKD):

1. Preliminaries	138,774
2. Access	80,000
3. Coating Lump sum, ducts	1,094,899
Provisional for Depot	4,442,055
4. Repair Lump sum, ducts	167,308
Provisional for Depot	2,733,200
5. Identification and recording	82,040
6. Contingency	300,00
Total	8,768,276

A 'Reviewing Engineer' engaged by ST (see 3) advised the Client on April 25, 1991 that he had been retained by ST to assist them in evaluating their contractual position, and requested a detailed cost breakdown. The Client confirmed that there was no change to the scope of repair work, except to the extent that some coating work would vary with site conditions.

ST's views were set out in a letter to AB dated June 25:

- The Reviewing Engineer, consulted as an expert, had advised ST that with a nominal cover of 15 mm, with 10 mm being acceptable (see tolerance under 2), corrosion to reinforcement within the relatively short period of five years was very likely. The damage, therefore, was due to a design fault; hence ST was not liable.
- ST would admit liability only for planks where the actual cover was less than 10 mm. It would consider application of protective coating to areas where cover was more than 10 mm, an element of betterment for which ST would not be responsible.
- The Reviewing Engineer had advised ST that the maximum liability thus would be limited to HKD 1,672,974.

On August 12 the Client appointed its own expert.

AB sought legal advice in September 1991 and was advised that

- the Client could bring an action for breach of contract within six years of the date of breach of contract;
- the Client considered the breach as having occurred at the date of substantial completion of works (September 16, 1985). If AB did not sign a limitation waiver agreement, the Client could commence arbitration;
- it was clear that AB was exposed to a claim in contract in respect of the defective pre-cast concrete planks.

Concerning AB's Claim against ST:

(1) In contract
- AB's contracts with ST were dated August 1982 and March 1984. The last delivery was in October 1984. The contract's clause 3(2) read as follows: 'Any claim relating to quantities of, or defects in, goods delivered must be made within 24 hours of delivery and confirmed in writing by the customer within 7 days of such delivery. In the absence of any such claims and confirmation, the goods shall be deemed to have been accepted and received in compliance with the customer's order.'

- The six-year period from the last date of delivery had expired, so any claim for breach of contract would be time barred; this position would have been different if ST had been fraudulent, but this was not the case.

(2) In Tort

In order to initiate an action for negligence, AB would have to show that ST owed AB a duty of care by manufacturing and delivering the planks, that ST had breached that duty, and that AB had suffered a loss as a result of ST's negligence. However, AB was not the owner of the defective goods and at the time in question had not suffered any loss. It therefore would appear that an action in tort would be difficult, and since ST was prepared to offer a cash settlement, that should be pursued.

In a letter to ST of September 12, AB
(a) accepted ST's offer of HKD 3 million towards a settlement with the Client, and
(b) stated that further discussion would be needed between AB and ST in respect of proportional liability for any amount greater than HKD 4 million.

20 Resolution

A negotiated settlement was arranged through letters and meetings between the Client, the Contractor and the Manufacturer/supplier.

Lawyers participated only through advice to the parties.

The cost of repair to be met by the Client was partially covered by contributions from AB and ST.

The Designer probably provided free services for producing documents for the repair works.

21 Costs

Each party probably met its own costs since the conflict resolution was mainly by negotiation between the parties, though experts were also engaged from time to time.

The Client subsequently claimed, and recovered, more than half the cost of the repairs from AB, who in turn claimed the major part of its outlay from ST.

22 Information

It appears from information obtained from one party that all the parties involved did not have access to all the others' information. This factor may have favoured a negotiated settlement.

23 Time

The Client attempted successfully to resolve the issue within the six-year period for action in respect of breach of contract. The period was calculated, in this case, from the date of substantial completion of the contract.

Interestingly, settlement was reached during the penultimate week of the period.

24 Role of Experts

As mentioned, ST engaged an independent expert (the Reviewing Engineer) and attempted to shift liability to the Designer.

All parties consulted lawyers.

The role played by insurers is not clear, but it may be assumed that they made some input to the process.

25 Outcome

As mentioned, a negotiated settlement was reached.

Both AB and ST acted with a 'long term vision,' not uncommon in negotiated settlements in Hong Kong. In fact, this case is illustrative of the local (and perhaps Eastern) preference for negotiated settlements in such scenarios.

On September 13, 1991, the Client accepted a lump sum of HKD 4.25 million from AB as a contribution towards the cost of remedial works.

On September 17, AB advised ST of an apportionment, as follows:
- ST's contribution would be HKD 3.125 Million (HKD 3 million plus 50 percent of HKD 250,000);
- AB's contribution would be HKD 1.125 Million (HKD 1 million, plus 50 percent of HKD 250,000).

Note: The original contract sum was about HKD 508 million.

No insurance settlement was involved in respect of AB or ST; it is not impossible that the balance of the repair cost was fully or partly met by the Client's insurance company.

30 Feedback

The case is not available for further reference, as the negotiations were conducted in private and the identities of the parties must be kept confidential.

Note: The case history is based on information made available by one of the parties. Authors, contributors and publishers accept no responsibility for details or interpretation in the foregoing.

Japan

Type A

Fumihiko Omori and Michiko Yamauchi

THE PURCHASER'S OPPORTUNITY FOR RECOVERY THROUGH LITIGATION OR OTHERWISE, FROM THE VENDOR

Relationship between the vendor and the purchaser

The relationship between the vendor and the purchaser in legal terms is determined by the provisions of either the Civil Law (CL) or the Commercial Law, unless there is a special exemption. In this case, the time of discovery of the cracks as well as the time of signing of the contract of sale would also affect the relationship.

1 The case that the cracks were manifest before the contract was complete.
1.1 The contractual responsibility is the first issue to be dealt with, and 'Defect Warranty' (CL Art. 570) upon completion of the contract of sale should be considered.

A 'concealed defect' is a defect which cannot be found through inspection at the usual level of attention and scrutiny applied to a normal transaction. If the cracks are considered as a concealed defect, a claim for damages or repudiation of the contract is possible. But this particular case cannot be categorised as one that revolves around a concealed defect, because the cracks should easily have been noticed by a normal site inspection. The Bank therefore might not be able to pursue the responsibility based on the contract with the vendor.

Though not applicable in this case, if the cracks affected the building in a functional way (such as having a negative effect on the performance of

Case Studies in Post-Construction Liability and Insurance, edited by Anthony Lavers. Published in 1999 by E & FN Spon, 11 New Fetter Lane, London EC4P 4EE, UK. ISBN: 0 419 24570 7

waterproofing), it could be conjectured that 'functionally' there is a 'concealed defect.'

If both the vendor and the purchaser can be considered as merchants in terms of the Commercial Law, Commercial Law could be applicable as a special provision of Civil Law (CL). Commercial Law Art. 526, Prov.1 stipulates that a purchaser (the Bank) should conduct an inspection without delay after the acceptance of the building, and give the vendor notice of any defect, also without delay; otherwise a purchaser (the Bank) could neither claim release from the contract nor compensation for damages. In this particular case, the Bank does not seem to have inspected the works, and the resolution would be the same as at Civil Law.

1.2 If some cracks existed at the time of purchase and the condition of the tiles deteriorated after the purchase, 'Nullity of the Contract by Reason of Mistake' (CL Art. 95) could be applicable. If the Bank could prove that it would not have bought the building if it had known that the tiles would deteriorate, and if the vendor somehow had been notified that the appearance of the building would not get worse than the *status quo*, the contract could become void because of mistake, and the Bank could claim the amount paid plus its loss (5 per cent p.a. of the amount paid). In this case, the Bank would be obliged to convey the building back to the vendor. If there were any significant negligence on the Bank's side, 'Nullity of the Contract through Mistake' would not be applied. For example, if significant cracks existed at the time of completion of the contract, the Bank would have limited rights to put forward a claim.

1.3 If the vendor did not know that the tiles would deteriorate, but this could have been foreseen by an appraisal of the situation, the purchaser could claim damages based on tort. In this particular case, because the data produced by the building research institute were available at the time of the transaction, the degree of difficulty in obtaining the data would nevertheless be a factor in determining the result.

The loss in this case includes the projected profit, but the damages recoverable would be limited by the causal relationship between the vendor's act and reasonable loss. Consequently, the purchaser could claim the amount equivalent to the reduction of the value, and, possibly, any capital gain obtained by resale.

2 The case if the cracks became discernible after the contract was complete.

2.1 The vendor assumes responsibility for defects under 'Defect Warranty' (CL Art. 570) which includes 'concealed faults'; therefore, the purchaser can claim rescission of the contract, or sue for damages. Since rescission of the contract could be claimed only when the purpose of the contract could not be fulfilled, the extent of cracking would become a point at issue. The purpose of the contract should be understood within the context of the nature of the property and other conditions that applied at the time of completion of the contract. Also, the decision that the purpose of the contract could not be fulfilled would be based on the fact that repair could not be performed easily and inexpensively.

Considering the fact that this case concerns a 28-unit business park, the appearance of the building could be one of the contractual factors. Therefore, if the property has an unsightly appearance due to the cracks, and it is not possible to repair them in an easy and inexpensive way, the contract could be repudiated. However, if the cracks could be repaired in an easy or inexpensive way, the purchaser cannot seek release from the contract and could sue only for damages.

The fact that the cracking is not of a magnitude that affects the stability and performance of the building would be a negative factor in trying to obtain the remedy of rescission. This would be the case if performance and stability criteria were proffered as the sole ground for seeking rescission.

However, if the purchaser could establish that the cracking was such that it would affect the potential to sell the building because of its unsightly appearance, this could be a highly persuasive ground for seeking legal remedy.

Once the contract is rescinded, both parties are obliged to restore matters to the *status quo ante*, and the Bank is entitled to claim the total amount paid as well as interest (5 per cent calculated retrospectively from the time of the completion of the contract). If there were other losses, such as the profit which could have been obtained if there had been no cracks, the Bank could also claim for these. A claim for lost interest could also be lodged.

The claim for rescission of the contract or compensation should be activated within one year after the Bank recognises the existence of the fault (CL Arts 570 and 566,3).

2.2 If the building's appearance was a very important factor for the Bank in reaching its decision to purchase, the Bank could possibly claim the amount of loss on top of the amount paid, based on 'Nullity of the Contract through Mistake' (CL Art.95), but this claim would be void if it were revealed that the Bank, had it made reasonable efforts, could have obtained the information regarding the cracks.

If it was foreseeable that the cracks would get worse after the contract was entered into, the purchaser could claim damages, alleging the vendor's negligence. The decision whether it was foreseeable or not would depend on how difficult it was for the vendor to obtain the information on the cracking.

THE PURCHASER'S OPPORTUNITY FOR RECOVERY, THROUGH LITIGATION OR OTHERWISE, FROM THE VENDOR'S CONTRACTOR

Since the Bank and the vendor's contractor do not have a direct contractual relationship, the only claim the Bank could possibly make against the contractor would be a claim for damages based on tort (CL Art. 709). Negligence by the contractor and the actual loss will be the point of dispute referred to above.

THE PURCHASER'S OPPORTUNITY FOR RECOVERY, THROUGH LITIGATION OR OTHERWISE, FROM THE VENDOR'S ARCHITECT

As the Bank and the architect do not have any direct contractual relationship (as above), tort (CL Art. 709) becomes a focal point at issue. But in this case the contractor would be less likely to be found negligent than the architect, because a contractor's responsibility in general is considered to carry out the work in accordance with drawings and specifications, and – as the architect's decision caused the fault, and given the high level of care that is expected of an architect –, the architect would be more vulnerable to a claim in negligence than the contractor.

In this particular case, a point at issue would be that the vendor insisted on the use of the tiles and that the information suggesting that such a product might be unsuitable was not widely circulated at the time of completion of the building contract. Indeed, the fact that the architect, despite uneasiness, acceded to the developer's wishes, could be a point at issue when the liability for negligence is apportioned between the vendor and the architect. The Bank would not be implicated in this particular inquiry.

Regarding the availability of the information, if the architect could prove that at that point in time it was impossible to obtain the information, even by professional and prudent scrutiny, it would be difficult to substantiate a claim in negligence.

THE INVOLVEMENT (IF ANY) OF INSURERS IN THE PURCHASER'S CLAIM FOR RECOVERY, INCLUDING THE METHOD OF RESOLVING THE MATTER

At present, there is no property insurance or 'Warranty Insurance' which would insure against this kind of defect in Japan. Though there is liability insurance for developers, this does not cover the loss caused to the building itself. The only situation in which insurance could be applied is where damages are claimed from an architect who is covered by an 'Architects' Liability Insurance'. Even in this case, the degree of the insurance company's involvement in a dispute settlement would be limited to giving some advice to the architect alleged to be negligent, and direct involvement would rarely happen because the 'Lawyer's Law of Japan' prohibits anybody except an attorney from acting as a legal representative in a case concerning Japanese liability insurance schemes, including 'Architects' Liability Insurance'. If the case affected the insurance company's interests, the company could take part in dispute resolution.

When an insured party needs a legal representative, e.g. in the case of a lawsuit, the insurance company, upon a request by the insured for consultation, may recommend a certain attorney, or may approve the selection of an attorney by the insured.

As mentioned, there are few cases where insurance companies get involved with dispute resolution directly. The insured architect would either assume conduct of the negotiations on his own or he might engage an attorney approved by the insurance company. If the architect intends to settle the problem privately, the architect should get approval in advance from the insurance company.

THE PROCEDURES BY WHICH THE ABOVE ARE RESOLVED, INCLUDING EXTRA-LEGAL, LEGAL AND INSURANCE PROCEDURES

1. Common dispute resolution procedures
1.1 The following are the general procedures for dispute resolution:

As described, if there is a possibility that the Bank could claim damages against any party, i.e. the vendor, the contractor, or the architect, recovery could be sought for the full amount of loss from each party on a joint-and-several liability basis. The total amount claimed, however, would be limited to the actual losses .

Practically speaking, the Bank would normally begin by negotiating with the vendor, who has a direct contractual relationship with the Bank. Such negotiations are not part of legal proceedings. The vendor would then bring the matter to the attention of both the contractor and the architect.

If the parties cannot come to an agreement by negotiation, the matter will be brought to conciliation or trial, both of which are part of legal proceedings.

While conciliation requires the agreement of the parties concerned, trial does not. In Japan, even when the case is proceeding to trial, most cases are resolved by conciliation, with a judge acting as a conciliator ('Trial by Conciliation'). If the case is not settled through conciliation, the last resort would be trial.

1.2 If it were decided that the vendor should take responsibility, the vendor would try to seek contributions from other implicated parties, such as the contractor and architect. The apportionment of responsibility between the parties would then become the subject of negotiation. If the vendor assumes a continuous business relationship with the construction company (the contractor), the vendor would, in most instances, take action against the contractor, who also would have a greater capacity to contribute financially, rather than against other producers, such as the architect. But most large Japanese construction companies (contractors) are extremely cautious when using new materials, and have very quick access to information on new materials, and the possibility of the occurrence of this kind of accident is very low.

Type B-I

Kouhei Matsumoto with Fumihiko Omori and Shuji Motoki

The *Mrs. T* case

1 The Plaintiff
The Plaintiff was an individual owner and occupier of a bungalow.

2 The Works
The exterior of the house was clad with siding boards.

3 Procurement Method
The house was built under a traditional design-and-build contract.

4 Insurance
The owner had no property insurance which would cover the damage.

The builder was covered by 'Liability Insurance to Warrant the Quality of Houses', provided by The Registration Organisation for Warranted Houses (ROWH), which registers builders and developers and carries out ongoing site inspection, akin to the NHBC in the UK. Twenty property insurance companies jointly offer this insurance cover.

5 Damage or Loss
Rainwater flowed through joints between the exterior siding boards, and stained the inside of the walls. The damage was discovered by the owner and occupier.

The error was use of inadequate materials and poor workmanship. According to the loss adjuster, some 40 per cent of the loss was due to materials and the remainder to poor workmanship.

The plaintiff requested repair of the exterior boards to stop the inflow of water and cleaning of the stained part of the house.

10 Process after Discovery
The owner (occupant) of the house claimed reparation from the builder.

11 Technical Reports
A claims report was prepared by the builder, at the request of the ROWH, the insurer, and at the builder's expense.

20 Resolution
The builder agreed to carry out the repair work and reconciliation was thus achieved.

21 Costs
The 'Liability Insurance to Warrant the Quality of House Construction', by which the builder was covered, paid a little less than 80 per cent of the total repair cost, and the builder paid the rest.

22 Information
At the assessment of the case by the insurer, both the builder and the house owner (occupant) are supposed to be present.

23 Time
It took one year and four months from the discovery of the damage to the completion of the repair works.

24 Role of Experts
No external experts were involved.

25 Outcome
See 20.

30 Feedback
The details of the case are not available to the public, as ROWH reveals only general information.

Type B-II

Kouhei Matsumoto with Fumihiko Omori and Ikuo Watanabe

The *Retail store* case

1 The Plaintiff
The plaintiff was the owner of a retail store.

2 The Works
The project was a retail store.

3 Procurement method
The plaintiff engaged a designer (an architect) and separate trades contractors.

4 Insurance
The designer carried Architect's Liability Insurance.

The client had no property insurance which would cover the damage (in Japan, there is no property insurance which would cover damage caused by subsidence).

5 Damage or Loss

The building's glass windows and metal-framed glass sliding doors were damaged by subsidence of the floor slab.

The damage was discovered by the shop-owner.

The owner lost profits by having to reduce the area of the shop which could be used during the repair works, in order to prevent further subsidence, and to allow repair of the damaged parts.

The error was in the design (insufficient subsoil exploration) and in workmanship (insufficient filling before concreting and rolling).

10 Process after Discovery

The shop owner initiated a claim against the designer and the contractor.

11 Technical report(s)

The plaintiff did not commission a report. The designer asked the party executing the repair work to conduct a test boring, and this was done.

20 Resolution

The plaintiff's legal action against the designer was not seen through, as an agreement was achieved before the court hearing. Likewise, the claim against the contractor was abandoned when conciliation was achieved.

21 Costs

The designer paid half of the lawyer's fee, the other half being paid by the Architect's Liability Insurance. Other costs were paid by each party.

22 Information

The test results prepared by the party executing the repair works, as well as other data, were made available to all parties concerned.

23 Time

The case between the plaintiff and the designer was settled one year and nine months after the damage was discovered; it took five years and three months to settle the dispute with the contractor.

24 Role of Experts

No experts (except the party executing the repair works) were engaged.

25 Outcome

Prior to the legal proceedings, the Dispute Resolution Committee (DRC) proposed the ratio of negligence between design and workmanship shown below. The DRC is established within the Association of Architects with support from the insurance companies, and could be consulted only by the member architects (Architect's Liability Insurance is also limited to the members of this Association). The DRC's resolution proposal is not normally announced in public. Since the ratio was proposed as a sort of reference by the DRC which technically represented the insurer, the insured could expect the approximate amount to be covered by the insurer. This figure would have helped the architect at the negotiation table.

The ratio proposed by the DRC was as follows:

60 per cent for poor design because of insufficient subsoil exploration.

40 per cent for poor workmanship because of back-filling before concreting and rolling.

The conciliation achieved through the Dispute Resolution Committee found JPY 14 million to be a reasonable amount to be paid to the plaintiff. The compensation for cases related to foundation problems is reduced in half, and the excess is JPY 1 million. The insurer thus paid (14 - 1) x 50 per cent = JPY 6.5 million, and the designer himself paid the remaining JPY 7.5 million.

The case between the plaintiff and the contractor was concluded by a conciliation whereby the contractor would pay JPY 15 million.

The loss in profits caused by the reduction of the area of the shop during the repair work was agreed to be at the plaintiff's own expense.

The authors have doubts about the fairness of this solution, because the plaintiff ended by paying JPY1 .5 million for repair, in addition to procedural costs, even though he had not committed any faults.

30 Feedback

The conflict between the plaintiff and the contractor was known to the public; such information is, in principle, open, though provisions for public reading are restricted in order to protect privacy.

Standardisation and regulatory bodies also provide feedback through an information booklet, but this contains only selected cases.

Kuwait

Type A

Jasem M. Jumaian

THE PURCHASER'S OPPORTUNITY FOR RECOVERY THROUGH LITIGATION OR OTHERWISE, FROM THE VENDOR

An *Ameeri Decree* (order issued by the Prince of Kuwait) was issued on November 1, 1980 clarifying a point of the Civil Law; this was published in the Official Gazette 'Kuwait Al Youm', issue No. 1335 of 5 January 1981. The answer to the question is to be found in the Civil Law, Vol.2, Ch.1, Sect.1, Cl.456, and depends on whether the purchase and sales contract provided for any warranty for the property after the sale. Generally, a warranty is not set forth in a purchase and sale contract. Nevertheless, as mentioned, a contract usually stipulates that the purchaser must inspect the property and accept it as it is.

The Bank can take the case to court if it is able to prove that the developer tried to cover the defective area (fraud) or mislead the inspection; if the Bank failed to prove this point, the judge would not consider the case. A relevant rule in Islamic law pertinent to this issue indicates that 'The contract is the law between the contracting parties.'

Case Studies in Post-Construction Liability and Insurance, edited by Anthony Lavers. Published in 1999 by E & FN Spon, 11 New Fetter Lane, London EC4P 4EE, UK. ISBN: 0 419 24570 7

THE PURCHASER'S OPPORTUNITY FOR RECOVERY THROUGH LITIGATION OR OTHERWISE, FROM THE VENDOR'S CONTRACTOR

The name of the owner, a description of the building and its location would be mentioned in any construction contract. This would maintain the contractor's responsibility for the building, even if ownership changes, so there is no need to transfer this responsibility from the former owner.

Kuwaiti Civil Law governs relations between the contractor and the owner. Referring to Vol.2, Ch.3, Sect.1, Cl.662, 663, 664, 665 and 666, and assuming that the contractor did not notice any defect on the cladding panels, the contractor will not be held liable.

THE PURCHASER'S OPPORTUNITY FOR RECOVERY, THROUGH LITIGATION OR OTHERWISE, FROM THE VENDOR'S ARCHITECT

The architect's responsibility is twofold: Firstly through the design process, and secondly by supervision during construction. Each stage entails responsibility.

• Design

Normally, in Kuwait, the designer is responsible for his design if the works are executed as per his specifications. If the contractor did not comply with the specifications, the contractor will be held responsible, as will a consultant responsible for the supervision, should a consultant accept such a supervision assignment. Clearly, if the contractor does not comply with the contract he will be held responsible.

• During construction (supervision)

The supervision consultant must inspect new materials and send samples to technical laboratories to carry out the tests required for approval. Alternatively, he should seek such information from the manufacturer.

THE INVOLVEMENT (IF ANY) OF INSURERS IN THE PURCHASER'S CLAIM FOR RECOVERING, INCLUDING THE METHOD OF RESOLVING MATTER

Most construction contracts oblige the contractor to insure the project during the construction period (All Risks and Third Party Liability). The insurance policy will generally cover damage to the works, other buildings, equipment and labour only during the period of site work.

Another factor affecting the insurance is the Islamic laws, which permit only to a limited extent insurance cover exceeding the amount of physical damage.

Also the architect is insured for his design: Professional Indemnity Insurance is legally required. This, of course, covers loss, damage, or both arising from professional negligence.

If the cover concerns third party insurance or if the claim is subject to a dispute, claims cannot be processed before a settlement through litigation has been reached and recovery from the insurer will not be paid without a court verdict. If there is no dispute and the cover is not for third party insurance, the insurer can pay to the beneficiary direct.

THE PROCEDURES BY WHICH THE ABOVE ARE RESOLVED, INCLUDING EXTRA-LEGAL, LEGAL AND INSURANCE PROCEDURES

In this case, the solution would probably be reached through negotiation between the Bank and the developer, and, presumably, the developer would agree to meet the cost of repair.

If this were not possible, we may assume that the Bank would file a lawsuit against the vendor. The court would then, first, refer a case like this to an Expert Department; this Department would inspect the building and assess the damage. This step could give the owner of the building the right to cancel the sales contract before the case was tried in court.

Once the inspection was done, and assuming that the sales contract was not cancelled, the Bank would decide to change the cladding in order to give the building a better appearance. To this end, it would consult an independent architect to propose a new cladding material, and then proceed to call for tenders. Thus, the costs involved would be known.

The Expert Department, after reviewing the case and considering the responsibility of each party, would probably conclude that each party did share a certain amount of responsibility.

According to Kuwaiti law and regulations, all parties must bear some of the liability. The following are some highlights of the responsibility borne by each party:

The developer would be held responsible because he insisted on that particular kind of cladding material. Therefore, he would be presumed to be aware of the suitability of the material, or to have achieved a level of confidence in the material by requesting a material specification certificate and conducting test before erection.

When the Bank contracted to purchase the building, it accepted to do so having in mind the condition of the building. A purchaser is legally presumed to have inspected the building he intends to buy, as is stated in normal sales contracts for real estate which contain a clause stipulating that the purchaser check the building and accept it as it is. Accordingly, it is obvious that the Bank made a mistake when it purchased the building without any inspection.

The court would probably uphold the Expert's advice.

The Bank, however, might challenge the verdict before the Court of Appeals.

ANY MECHANISMS EXISTING FOR UTILISING THE FEEDBACK OR EXPERIENCE FROM THE CASE TO A BODY OR BODIES WHICH MAY MAKE USE OF IT

None.

Type B–I

Jasem M. Jumaian with Warba Insurance Co. S.A.K., with special thanks to Mr. Tawfiq Al Bahar, General Manager

The *Global Builders* case (fictitious name)

1 The Plaintiff
As there was no court case, the term 'claimant' is used in the following instead of 'plaintiff'.

The claimant was a highly reputed building contractor with several years' experience who lodged a claim with an insurance company.

2 The Works

The contract was for 352 houses to be handed over in two stages. The client for the project was a national housing authority, hence in the public sector.

The buildings were standard two-storey bungalows of less than 300 sq.m each, built of rendered block walls under a flat reinforced concrete slab roof with the cladding being of sand-lime bricks around aluminium window frames.

The period for completion was two years with a one-year maintenance period for civil works and two years for electrical and mechanical works.

The total value (in 1984) was some KWD 13 million.

3 Procurement Method

The contract was a typical 'Design-and-Build', with separate trades.

4 Insurance

The contractor took out Damage Insurance and Contractors All Risks Insurance. It is assumed that the designer carried liability (Professional Indemnity) insurance.

5 Damage or Loss

The damage was observed by the contractor.

The following is a quotation from a Surveyor's Report:

> We are informed that the problem with regard to the sand-lime cladding bricks [in the façade, anchored to the inner wall by steel wire butterflies] started during construction, when the Contractor(s) observed that bricks were falling down from various levels and/or the panels of bricks were moving laterally at some places or cracks started to appear here and there in the facia works.

While the buildings were left unoccupied (below), further damage ensued. The Loss Adjusters made the following remarks:

> Obviously the condition of the buildings had deteriorated, due to the fact that they had been standing unoccupied. The insurer's security arrangements were negligible and occupants of houses in adjoining areas had used the building [site] as a source of manhole covers, electrical switches, sanitary ware etc. for their unofficial extensions. Most of the windows not already removed had been broken and sand was drifting into the buildings. Damage to the outer leaf of brickwork seemed to be universal and was found in all areas of the Project.

Also a personal injury aspect was mentioned by the Loss Adjusters:

> The detail of the parapet walls - a free-standing masonry wall with reinforced concrete coping - was quite simply a death-trap. These wall were not capable of withstanding any intense horizontal loading and when they had undergone some deterioration, a strong gust of wind could have brought them down on a passer-by.

10 Process after Discovery

The contractor made a claim for KWD 1 million to his insurer. On the basis of surveyor's reports (below), the claim was repudiated because the policy excluded loss or damage due to fault, defect, error, failure or omission in design, plans or specifications. This happened in 1986.

In 1989, the case was reopened. The insured now lodged a claim for KWD 697,341. This time, the insured claimed that the damage was caused by bad workmanship.

The client stopped payments; the contractors then sued the client who counter-sued, upon which the contractors withdrew their case and agreed in principle to carry out the repairs on their account.

The buildings then were left unoccupied, and further damage ensued (above).

11 Technical Reports

Technical studies were conducted by two separate organisations. In brief, the reports said that the design of such walls must accommodate permissible tolerances as recognised from practice. Thus, both reports stated that the problem of sand-lime bricks falling was due to inadequate design, and that the design and the details as shown by the Principals did not meet the building codes and were not in conformity with standard practice.

The independent investigators' reports referred to above, and a Ministry of Justice Expert's report commissioned following the initiation of (the aborted) legal process between the insured and the client, confirmed that the client's design incorporated many unfortunate details, creating inevitable problems with the outer brickwork, possibly because the ties between the walls were designed for typical United Kingdom type cavity walling.

The reports recommended the replacement of the walls by rebuilding 110 mm thick walls resting on steel angles anchored to the ground beam round the building, with the top end of the wall fixed to a tie beam and properly sealed.

20 Resolution

Under the material damage section of the policy, loss or damage due to defective design is excluded (Exclusion 1). Similarly, the costs of rectifying or replacing defective material and/or workmanship is excluded (Exclusion 2); however, other work or damage to property as consequence of such defect is covered. Exclusion 2 is 'limited to the structural or work directly affected and shall not extend to other work or property lost or damaged as a consequence of such defective materials and/or workmanship.'

The claim was settled on a compromise basis.

21 Costs

The cost of the reports was covered by the client and the contractors.

22 Information

The information summarised above was available to all parties.

23 Time

Not known.

24 Role of Experts

The parties used their own experts.

25 Outcome

The compromise settlement amounted to KWD 50,000.

30 Feedback

Insurance claims are strictly confidential and published only if they go to court.

Norway

Type A

Jan Einar Barbo

THE PURCHASER'S OPPORTUNITY FOR RECOVERY, THROUGH LITIGATION OR OTHERWISE, FROM THE VENDOR

1.1 INTRODUCTION

The contract between the investment bank (the purchaser) and the developer (the vendor) is a contract for the sale of real estate. Thus, the Sale of Real Estate Act of 1992, *lov om avhending av fast eigedom 3 juli 1992 nr.93*, will apply.

The parties may, however, contractually have agreed upon conditions which differ from these statutory provisions, for, as a general rule, the *Act* permits derogation, save in contracts where the buyer purchases the real estate for his or her personal use (e.g. for residential purposes). In the case, however, no information is given as to the contents of the contract between the parties. It is therefore presumed that the contract is in accordance with the *Act* (for example, no exemption clauses etc.).

The case then raises three questions:

- whether the unsuitability of the cladding tiles constituted a defect which makes the vendor in principle liable for breach of contract (1.2 below);
- whether the purchaser's lack of inspection of the building precludes it from pleading that this defect constituted a breach of contract (1.3 below); and
- if the purchaser has a valid claim against the vendor, the amount of damages claimable (1.4 below).

Case Studies in Post-Construction Liability and Insurance, edited by Anthony Lavers. Published in 1999 by E & FN Spon, 11 New Fetter Lane, London EC4P 4EE, UK. ISBN: 0 419 24570 7

1.2 Is there a defect for which the vendor in principle is liable?

The answer to this question should be quite clear: According to the Act quoted, Sect. 3–2 (1), the property has to be suitable for the purposes for which such properties *usually* are intended, and also for any *specific* intended purpose, if the vendor was or ought to have been aware of this purpose (in effect, 'fit for its purpose'). If the property does not conform with any of these requirements, there is a defect for which the vendor in principle is liable.

It should be pointed out that there might be a defect (at the time the risk passes from the vendor to the purchaser) even if the defect materialises in damage at a later stage (*Act* Sect. 3–1(2)). It should also be pointed out that the vendor might be liable for breach of contract even if it acted in good faith: the liability for defects is strict. However, lack of negligence on the part of the vendor will have a major impact on the amount claimable by the purchaser (see 1.4 below).

In the case under scrutiny, it follows from the facts that the superficial cracks, and the subsequent poor aesthetic appearance of the building, made it unsuitable for the intended purpose, which was to let the units. The defect (the unsuitability of the tiles) was present at the crucial time (when the risk passed from the vendor to the purchaser – as a general rule, the moment at which the purchaser took over the use of the building), and consequently the vendor is liable for breach of contract.

1.3 The significance of the purchaser's lack of inspection

According to the *Act* Sect. 3–10, the purchaser cannot plead that a defect constituted a breach of contract if the purchaser was or ought to have been aware of the defect at the time the contract was made. It follows from the facts of the case that the purchaser was not aware of the defect, nor of the risk that this kind of tile might deteriorate.

The same provision also establishes that the purchaser cannot plead that a defect constitutes a breach of contract if it has inspected the property before making the contract, or without a sound reason has not complied with the vendor's request to make such an inspection, and such an inspection should have made it aware of the defect. (This provision does not apply if the vendor has acted in bad faith.) It is crucial that the vendor has to *encourage* the purchaser to make such an inspection: the purchaser is, according to the wording of this provision, not obliged to make such an inspection without the vendor's request. This interpretation is, however, somewhat disputed in Norwegian legal doctrine. Anyway, it is clear that *if* there is a duty to make an inspection without a request from the vendor, this inspection may be only superficial.

If the cracking of the tiles was undisguised at the time the sales contract was made, it seems clear that the defect cannot be pleaded (it is unlikely that the purchaser had not seen the building at all when the contract was made). If, however, the cracking could only be discovered by a careful inspection, e.g. only by experts (which in this case seems more likely), it is quite clear that the lack of inspection does not prevent the purchaser from claiming that the defect constituted a breach of contract.

1.4 The amount of damages claimable

Where there are defects, the purchaser may claim a price reduction (*Act* Sect. 41–12(1)). The reduction will in principle be *proportional*, that is, the contract sum will be reduced by a percentage equal to the percentage reduction of value of the property which the defect has caused. The consequence is that the price reduction will vary from the reduction in value only if the contract price was not equal to the property's market price. Unless another amount is proved to be appropriate, the price reduction will be set at the same amount as the cost of remedying the defect (*Act* Sect. 4–12 (2)).

The purchaser may also claim damages for losses incurred by the defect which are not covered by the price reduction. However, the vendor is only liable for so-called indirect losses if it has acted negligently (*Act* Sect. 4–14; indirect losses are, *inter alia*, losses due to reduced production or turnover and losses due to reduced possibilities of using the property in accordance with the intended purposes (*Act* Sect. 7–1(2)a and b)). Here, therefore, the purchaser will not be able to claim damages for loss of profit due to the impossibility of letting the units in the building.

THE PURCHASER'S RESPONSIBILITY FOR RECOVERY, THROUGH LITIGATION OR OTHERWISE FROM THE VENDOR'S CONTRACTOR

In principle, a subsequent owner may use its remedies for breach of contract because of a defect against a previous vendor or other contractual party, e.g. the vendor's contractor or architect, as in this case (*Act* Sect. 4–16(1)). This principle is now laid down in most modern Norwegian legislation concerning contracts, and is also gaining a foothold in modern standard form contracts. It is a condition that the vendor could have made the same claim against the party with whom it previously had a contract, and such a direct claim must be submitted before the defects liability period in the previous contract lapses, cf. the *Act* Sect. 4–16 (2). This is because of the subrogation principle upon which this provision is based. Thus, if the construction contract implements the Standard Form of Contract NS 3430 (which is the most commonly used standard form in major projects), the purchaser's claim has to be submitted within three years from delivery to the vendor.

It follows that the purchaser's claim against the contractor will – in this case – depend upon whether the unsuitability of the cladding tiles constituted a breach of the construction contract. It is also a condition that the unsuitability of the tiles represented a defect in the sales contract, and any liability on the part of the contractor will be limited to the amount which the purchaser could have claimed from the vendor.

There is no doubt that the contractor was not in breach of contract (it is presumed that there were no errors or omissions in the execution of the construction contract): the cause of the defect is in the design, for which the contractor is not responsible. The contractor would only be liable (wholly or partially) if it ought to have understood that the design was faulty, and there are no indications in this direction. The purchaser will therefore not be able to seek recovery from the contractor.

THE PURCHASER'S OPPORTUNITY FOR RECOVERY, THROUGH LITIGATION OR OTHERWISE, FROM THE VENDOR'S ARCHITECT

The same rules will apply in the relationship between the purchaser and the architect as in the relationship between the purchaser and the contractor. Thus the purchaser may be able to claim damages from the architect only if the vendor could do so. If the contract is based upon the Standard Form of Contract NS 3403 (which is currently being redrafted and until recently was the most commonly used standard form in client-architect relationships), the liability period for the architect is three years from delivery of the completed building, and subsequently the purchaser's claim would have to be submitted before this period lapses.

The vendor could only have claimed damages from the architect if the architect had acted negligently in the execution of his duty to design the building, or in his duty to advise his client (the vendor). The facts of the case indicate that the vendor obliged the architect to use this particular product for the cladding tiles, that the architect protested and specifically stated that he had no knowledge of the product, and that the developer nevertheless insisted. In these circumstances, it is submitted that the developer has to bear the risk of the product being unsuitable, especially in the light of this company having experience of commercial development projects. It follows from NS 3403 Cl.12.2 that the architect will not be liable if the client has approved a certain solution and the architect has notified the client of the possible risks. If the information about this product's possible tendency to crack had been widely circulated at the time the design was made, there might have been liability on the part of the architect if he were not aware of this, and therefore did not warn the client against this specific risk. However, the relevant information was not common knowledge in the industry prior to design and construction. Thus, this development risk would have to be borne by the vendor, and the architect cannot be blamed for having acted negligently.

If the architect were held liable and the contract was based upon NS 3403, the liability would have been limited to NOK 1,500,000 according to NS 3403, Cl.12.1.

THE INVOLVEMENT (IF ANY) OF INSURERS IN THE PURCHASER'S CLAIM FOR RECOVERY, INCLUDING THE METHOD OF RESOLVING THE MATTER

Neither the purchaser's property insurance nor the vendor's third party liability insurance will cover damage to the building arising from errors in design or construction.

The contractor is not liable for the defect (above), and even if it were, its Contractor's All Risk insurance would not cover this liability, as it would be arising from a contractual obligation only. The architect's possible liability, however, would be covered by his professional indemnity insurance, with (most likely) a limitation to NOK 1,500,000. The purchaser would have a direct action claim against the insurer, according to the Insurance Contracts Act 1989 (*lov om forsikringsavtaler, 16 juni 1989, nr.69*) Sect. 7–6. It is likely that the insurer would handle the claim in co-operation with the architect.

Recently (1997), a German insurance company introduced a so-called 'BUILD-insurance' through a Norwegian subsidiary. This covers costs of remedying defects for 10 years after hand-over. A condition is that the defect constituted a breach of contract on the part of one of the producers (contractors, suppliers, architects, engineers, etc.): there is no insurance cover if the owner bears the risk of the defect himself. This insurance 'follows the building', i.e. also subsequent owners are covered (during the ten year-period). Since the vendor (the developer) bears the risk of the cladding tiles' suitability, this insurance would not provide cover. If the contractor or the architect had been liable for the defect, however, the purchaser would have been able to seek recovery from the insurance company (which then could have recourse against the producer responsible, if the claim against the producer was submitted within the defects liability period).

THE PROCEDURES BY WHICH THE ABOVE ARE RESOLVED, INCLUDING EXTRA LEGAL, LEGAL AND INSURANCE PROCEDURES

After the purchaser discovers the cracking of the tiles (and the impossibility of letting the units in the building), there will first of all be investigations in order to disclose the cause of the damage to the tiles and to arrive at a decision as to how to remedy the damage in the most efficient way. Depending on the purchaser's financial situation, it will most probably replace the tiles as soon as possible, in order to mitigate the economic loss; the purchaser would immediately engage building engineers to this end.

At the same time, or after the disclosure of the relevant facts, the purchaser will engage either an in-house or an external lawyer to deal with the case. Then a formal claim against the vendor will be submitted. However, it is likely that the vendor will already have been notified, albeit without a precise claim from the purchaser.

If the purchaser chooses to make a claim against the architect, this will probably be submitted at the same time. When the architect receives the claim, he will immediately contact his insurer, and the insurer will to a large extent handle the claim in co-operation with the architect, who would be likely to engage a lawyer as well. There will most probably not be any other insurers on the scene (above), and the purchaser, if properly advised, will probably not submit a formal claim against the contractor.

At this stage, there will be intensive negotiations. All parties will try to avoid litigation. Litigation is costly, it consumes the parties' resources, and it takes time. Statistically, these negotiations are most likely to end in an amicable settlement.

If, however, the parties do not reach an agreement, the case would probably go to court. Alternative dispute resolution, e.g. arbitration, is most common in contractual relationships in projects of a technical nature. In a relatively clear-cut case like this — which first of all seems to raise questions of law, not facts — arbitration or other alternative dispute resolution procedures are not very widespread.

If the purchaser chooses to go to litigation, the case will commence in the City or County Court which covers the area where this particular real estate is located (there are some 90 such courts in Norway). This court is presided over by one judge. As the date for the hearing comes closer, the possibilities of reaching a settlement out of court will increase. In fact, many cases are settled during the last few days before the hearing, because the parties at this stage become more conscious of the possible risks.

The period of time from the purchaser's issuing the writ to the City or County Court's decision will usually be between six months and up to three years. The time will, of course, depend upon the pressure of work in the court, and to what extent the parties keep on with their negotiations during this period.

There may be an appeal following the City or County Court's decision (there are six Appeal Courts in Norway). Without permission from the Appeal Court's chairman, an appeal cannot be heard unless the claim in question involves an economic value of at least NOK 20,000. The Appeal Court is presided over by three judges. Its decision will usually be reached within one or two years after the decision of the inferior court.

There may be an appeal to the Supreme Court following a decision in the Appeal Court. If the claim in question involves economic values of less than NOK 100,000, a special committee in the Supreme Court will have to give permission for the case to be heard. It will only give such permission if the further appeal involves questions of a fundamental nature, or if the case involves questions of special importance to the appellant. The same committee may, regardless of the amount in question, refuse to let a case be heard for a number of reasons. Relatively speaking, very few cases are squeezed through this 'eye of a needle'. The practical consequence is that the Appeal Court's decision will be binding and final. If the appeal is accepted for a hearing by the Supreme Court, there will be five judges sitting. No witnesses or experts will appear before the court, only the parties' attorneys. The hearing is based upon documentary evidence and oral pleadings from the parties' counsel. The Supreme Court's decision will most likely be reached within two or three years after the Appeal Court's decision.

ANY MECHANISMS EXISTING FOR UTILISING THE FEED-BACK OR EXPERIENCE FROM THE CASE TO A BODY OR BODIES WHICH MAY MAKE USE OF IT

In Norway, there are no formalised systems for utilising the experience which participants in the construction industry obtain from discovering defects occurring in their projects. However, the Norwegian construction industry is relatively small (compared to larger jurisdictions), so in fact bad experiences with certain products will usually be well known in the industry after a while, at least if the experience is gained from one of the larger projects or by one of the larger firms in the industry.

Type B–I

Jan Einar Barbo with Hans Jakob Urbye

The *Fjeldhammer Brug* case

1 The Plaintiff
The plaintiff was a developer, Thorbjørn J Nordby (formally a master builder, because he did not have authorisation to act as a building contractor), erecting two seven-story blocks of flats.

2 The Works
The blocks were built on a slope. Each flat had its own terrace which constituted part of the ceiling and roof of the flat underneath. The terraces had a low parapet and a slight gradient, so that water would flow towards the drainage in the parapets between the flats.

Work on the roofs and terraces was executed during 1966 and finished in December the same year.

3 Procurement Method
The developer (hereafter 'Nordby') engaged the construction firm A/S Fjeldhammer Brug (hereafter 'FB'), a roofing expert, to do the roofing and the waterproofing of the terraces. The contract was based on FB's standard conditions, printed on the back of its offer. FB guaranteed that the terraces

would be waterproof and that there would be no condensation. The technique to be used was chosen by FB.

4 Insurance
Not applicable.

5 Damage or loss
In January 1967, Nordby submitted a written notice of defect to FB where he pointed out in general terms that the work done did not seem to comply with the quality he had been led to expect.

Throughout the spring of 1967, major leakages occurred, with water penetrating into the flats.

It proved extremely difficult to find the cause – or causes – of the leakages. Even five years later (at the time the Supreme Court pronounced its judgment, below), it was still not possible to be sure. Suffice it to say that there were several causes, varying from one terrace to another: on some terraces the leakage was caused by faulty design by FB, and on some by work executed by Nordby himself.

The leakages led to expenses because the waterproofing of the terraces had to be redone, and to a substantial loss because the whole project was delayed, as Nordby could not use the buildings for their purpose until the leakage problems were solved.

10 Process after Discovery
To Nordby's notice of January 1967, FB replied that the terraces were waterproof at the time of delivery (December 1966), and that no leakage was caused by its work.

Even after the leakages became obvious, FB still did not accept any liability for the damage, and argued that the cause of the leakages was Nordby's own works, not those of FB.

It must have seemed clear to Nordby that it would not be possible to reach an amicable settlement with FB. Consequently, he sued FB in May 1967, a few months after the first leakages had occurred. The Court was asked (by Nordby as the plaintiff) to appoint technically expert judges.

FB appealed and Nordby cross-appealed to a higher court. Finally, FB appealed to the Supreme Court.

11 Technical Reports
Both parties engaged technical experts at an early stage and presented a number of technical reports, and the experts acted as expert witnesses during the trials as well. They had fundamentally differing opinions on the

technical issues, which, of course, made the basis for an assessment of the proofs even more difficult. Thus, the Supreme Court (below) had to rely partly on the principles of burden of proof to establish the evidence.

20 Resolution

A City or County Court (in this case Oslo City Court) is usually presided over by one (professional) judge (save in criminal cases, where the court always consists of one professional and two lay judges); however, in (civil) cases involving technical questions, e.g. construction disputes, the court may appoint two more judges who are experts on the relevant technical issues and not lawyers (the three judges have the same decisive authority on both facts and law. Thus it is possible for the professional judge to be in the minority, even on questions of law; however, the professional judge will in practice have a major influence on the lay judges when it comes to purely legal questions).

Oslo City Court reached its decision in May 1968, one year after the writ was issued.

FB appealed to *Eidsivating lagmannsrett*, one of the five appeal courts in Norway at that time (now there are six). However, during the summer and autumn of 1968, FB did some of its work on the terraces over again, but reserved the right to claim compensation from Nordby in case the repair work exceeded what the Appeal Court would find FB obliged to do. As mentioned, Nordby cross-appealed, pleading that the damages awarded were too small.

An appeal court is usually presided over by three (professional) judges. In this case, another four technically expert judges were appointed, according to the same principles as in the City Court. The Court reached its decision in March 1970, almost two years after the City Court's judgment was pronounced.

FB appealed to the Supreme Court.

The Norwegian Supreme Court consists of 19 judges, but plenary hearings are ver000000000y rare. Most cases are heard in one of the divisions, presided over by five judges. (There are two divisions, and the composition of the divisions is based upon a principle of rotation.) No witnesses are examined in the Supreme Court. The hearings are based on documentary evidence and oral pleadings from the parties' counsel. Not even the parties themselves appear before the court (though they may be present, as may the public at large). Because the evidence has to be documentary, there will often be a hearing of the evidence in the City Court, and the documents from this hearing will be submitted to the Supreme Court. Thus, in the Supreme Court, recitation of statements from witnesses will replace oral examination. In this case, representatives of the parties and 18 witnesses and expert witnesses testified.

The Supreme Court pronounced its judgment in May 1972, two years after the Appeal Court's decision.

21 Costs

In accordance with the main rule in Norwegian civil procedure law, Oslo City Court awarded Nordby legal costs (lawyers' fees; expenses for obtaining expert opinions and other costs).

At the appeal, Nordby was also awarded legal costs incurred at both the City Court and the Appeal Court.

In the Supreme Court, FB's appeal was partly successful, but most of the legal costs were connected with the handling of the liability question, which in economic terms was lost by FB (below). FB was therefore ordered to compensate Nordby for parts of his legal costs for all the three courts.

22 Information

See above.

23 Time

As mentioned, five years elapsed from the first occurrence of damage to the Supreme Court's verdict.

24 Role of Experts

As already mentioned, both parties engaged experts who also acted as witnesses in the two lower courts.

25 Outcome

Oslo City Court ordered FB to redo all the contractual work. Nordby was also awarded a substantial amount in damages for various kinds of economic loss because of FB's breach of its duty to repair the terraces.

The Appeal Court held that FB was obliged to repair all the terraces, not only those it had already repaired. The judgment specified in detail the repair works FB was to execute. Nordby was also awarded damages, as in the City Court, but with a significantly higher amount. There were several dissenting opinions, the majority consisting of one of the professional judges and three of the technical expert judges.

The Supreme Court only partially agreed with the inferior courts. In its general remarks, it held that the warranty given by FB (above) should not be interpreted so that FB was liable for *any* leakage: it had not issued a 'water-proofing warranty' covering all kinds of leakage, no matter what the cause. It should be interpreted so that FB had a duty to repair defects caused by its own design, techniques, products, materials and works, and defects caused by those of Nordby's own works executed in accordance with FB's instructions. FB should also be liable for any consequential losses suffered by Nordby, including losses because of delay caused by these defects.

Having established these principles, the Supreme Court held that FB did have a duty to repair three of the terraces it had already repaired (in 1968, after the City Court's decision). However, the cause of leakage on a fourth terrace, which had been repaired in 1968, could be traced to Nordby's own works, including works executed after FB's works had been handed over to Nordby. FB was therefore awarded compensation for the repair works done on this terrace. Leakage on other terraces, which had not been repaired by FB, were also – most likely – a consequence of Nordby's own works. The basic conclusion was that FB in 1968 had fulfilled its obligation to remedy defects for which it was liable, save a few minor details for which FB accepted liability.

The greatest importance of the case, however, lies in the Supreme Court's handling of Nordby's claim for damages. He had suffered a substantial loss because the project was delayed due to the infiltration of water. There was no doubt that FB was liable for the consequential losses caused by defects covered by the warranty.

FB argued that it was not liable for delay caused by leakages not covered by its warranty. In principle, this is (of course) right. The Supreme Court, however, held that FB had a duty to investigate the problems when Nordby submitted his notice of defect, and to execute the necessary repair works demanded by Nordby, unless the investigations clearly showed that the leakages were caused by Nordby. FB was the expert, and it would have been difficult for Nordby to find another contractor to do these works. The cost of remedying would not have been substantial compared with the significant loss Nordby would suffer if the project was delayed, and FB was aware of this.

It will be noted that the Supreme Court to a large extent required FB to act in accordance with the principle of good faith and fair dealing, even after completion of its work. Even though it was established in the Supreme Court's decision that FB was not liable for the leakages on a number of the terraces, FB was nevertheless in breach of a contractual duty when it failed to remedy the leakages on those terraces. FB should have done so, if necessary with the reservation that it was entitled to compensation, and, if FB so wished, subject to a surety bond from Nordby. (It may be added that the Swedish Standard Form of Contract *AB92*, Cl. 7:24 codifies the principle that the contractor has a contractual duty to remedy disputed defects, even if it is later established that it was not liable for these defects.)

Because of FB's breach of this duty, it was in principle liable for all consequential losses caused by the delay: Nordby's loss of income from rents, increased financial costs, labour, site and plant expenses and loss of profit and fixed overhead expenditure during the period. However, Nordby had to bear one third of this loss, primarily because he had breached his duty to mitigate the loss: the Supreme Court found that Nordby had not put enough energy into trying to avoid the delay.

30 Feedback

The case is reported in the Norwegian Supreme Court Reports, *Norsk Retstidende*, Rt. 1972 p.449.

This case is now some 25 years old, but the result would probably be the same if the case arose today. The Supreme Court's reasoning, however, would probably be based upon a more explicit reference to the requirement of good faith and fair dealing in the performance of contracts. This principle was not that clear in 1972, but has been heavily accentuated in case law over the last 10–15 years. It is submitted that the *Fjeldhammer Brug* case is a precedent for the principle that the requirement of good faith and fair dealing not only applies during the performance of the contract, but also during the subsequent defects liability period.

Type B–II

Jan Einar Barbo

The *Bechs* case

1 The Plaintiff

The plaintiffs were Mr and Mrs Bech, a household.

2 The Works

The works consisted in a small annex to the Bechs' house: an extension of the living room and a covered terrace.

The foundation for the annex was piling, a foundation method chosen by the contractor.

The works were carried out at the end of 1991, and the Bechs paid the agreed contract price, NOK 75,000.

3 Procurement Method
The Bechs engaged a contractor (a one-man building firm).

4 Insurance
Not applicable.

5 Damage or Loss
During the winter 1992–93, the Bechs found that the annex had sunk several centimetres. The walls had cracked and water penetrated into the house; the ceiling, floor and wallpaper were damaged.

After an oral notice (below), the contractor jacked up the annex. However, the same defect and damage occurred during the summer of 1993, and the annex was once again jacked up by the contractor; the same damage occurred again in 1994, this time with no remedial intervention by the contractor.

According to a technical report made by another contractor in January 1995, the soil conditions on the site required that a concrete foundation for the piles should have been made to prevent them from sinking. The report also found that some carpentry work was sub-standard.

The estimate for remedying the foundations, the poor carpentry and the consequential damage was NOK 36,700.

10 Process after Discovery
An oral notice of defect was submitted to the contractor immediately (the contract was not based upon a standard form, and the notice was not required to be in writing).

After the jacking up in 1993, the contractor refused to execute any more remedial works, arguing that the cause of the ongoing subsidence was the client's lack of filling in soil around the piles in due time. However, the technical report mentioned above held that the absence of in-fill was of no importance to the subsidence.

In October 1994, the Bechs made a formal complaint through the Norwegian Consumer Council, *Forbrukerrådet*. The Consumer Council is an independent body financed by the Government. It has numerous tasks, among them handling claims from consumers concerning goods and services. In its claims handling role, the Consumer Council is not supposed to be a deputy for the consumer: it must act impartially and try to mediate between the parties in order to reach an amicable settlement. It has no authority to make binding decisions.

However, the contractor did not even reply to the Council's letters. Consequently, no settlement could be reached, and the Consumer Council closed the mediation by notice to the parties in February 1995.

The Bechs then demanded that the case be heard by the Consumer Disputes Committee (CDC), *Forbrukertvistutvalget.* CDC is a court-like body, financed by the Government and wholly independent in its judicial decisions. It tries cases where the parties have not reached a settlement through the Consumer Council's mediation, provided that one of the parties – within four weeks after the Consumer Council has closed the mediation attempt – requests that the case be heard by the CDC. Furthermore, it only has authority to try cases which concern sale of goods, provided that the goods are purchased for the buyer's personal use, and cases which are governed by the Supply of Consumer Services Act 1989, *lov om hand-verkertjenester m.m. for forbrukere, 16 juni 1989 nr. 63.* This act (in broad terms) governs contracts for the supply of services on (movable) goods, as well as services on real estate, except erection of new houses. The act applies only if the supplier acts in the course of a business, and only when the services are for the other party's personal use. The present case therefore was within CDC's authority, as it did not concern the erection of a new house but a small extension to an existing house.

CDC has 11 members: one chairman, two vice-chairmen, and eight other members. The three chairmen must be lawyers. Each case is heard by one of the chairmen and two other members (the author being the presiding chairman in the present case).

There is no oral hearing, and the parties are not present: the CDC's decisions are based upon documentary evidence only.

There is no charge for bringing a dispute before the CDC, but, on the other hand, CDC has a very limited authority to award legal costs, though expenses due to the procurement of technical reports, expert opinions and similar expenses may be awarded as ordinary damages.

CDC hears more than 500 disputes annually. Because of litigation costs in ordinary courts, it is to all intents and purposes the only body which makes binding decisions on small consumer claims. The present case is not necessarily representative in terms of size of claim. Most disputes concern claims up to NOK 20,000, with many of them being for not more than a few hundred NOK. However, there is, in principle, no limitation to the CDC's formal authority as to the amount (or other remedy) claimed, but larger claims are rarely heard by the committee, for two reasons: First of all, most consumers (and their legal advisers, if any) are probably not aware that the CDC has authority to handle even the bigger cases, and even if they were aware of this, ordinary litigation is probably considered a more satisfactory procedure because of the lack of opportunity to hear witnesses and parties in the CDC; secondly, the CDC has a wide discretion to refuse to hear a case if it considers the documentary evidence to be unsatisfactory as a basis for assessment of proof. In cases involving large claims, this will often be the case. But the CDC has made decisions in cases concerning claims of several hundred thousand NOK.

The parties may, if they wish, be represented by an attorney or other agent during the proceedings (but not during the formal hearing, cf. above). However, the opportunity to have an attorney is rarely used, and was not used in the present case. Legal costs would usually be disproportionate to the claim, especially in the light of CDC's limited authority to enforce payment of such costs.

CDC's decisions, which are public, are given with written reasoning, much like ordinary judgments by courts, though usually not as extensive. They are binding, and have the same legal effect as judgments by ordinary courts, unless one of the parties issues a writ in the City or County Court within four weeks after being notified of CDC's decision. It may be said that CDC to a large extent acts like a court, with the City or County Court as an 'appeal court'. Therefore, CDC is often referred to as the 'Consumer Court'.

After preparing the case for the hearing, where the parties were given the opportunity to submit further statements and written evidence, the case was tried in a CDC meeting in September 1995, and the written decision with reasoning was given the next month. None of the parties brought the dispute to the City or County Court, so the decision is binding and final, and may be enforced in the same way as an ordinary judgment of a court.

20 Resolution

CDC relied heavily upon the technical report made in January 1995, and found that the lack of a foundation underneath the piles most certainly was the defect, and consequently the cause of the damage. Since the contractor had chosen the method of foundation, this was undoubtedly a defect for which he was liable according to the Supply of Consumer Services Act 1989 Sects 5 and 17. The Bechs were therefore entitled to a price reduction, which, according to the Supply of Consumer Services Act 1989 Sect. 25(2), should be the cost of remedying the defect.

The supplier of services is, according to the Supply of Consumer Services Act 1989 Sect. 28(3), liable for damages for consequential losses if it does not prove that the damage is not caused by negligence (reversed burden of proof). The contractor had not even tried to submit such evidence, presumably because he in fact agreed with the technical report. The Bechs therefore were entitled also to compensation for the consequential losses caused by damage to the ceiling, floor and wallpaper.

21 Costs
The technical report from the contractor mentioned above was made on the Consumer Council's request, but paid for by the Bechs (but see below). The Bechs had not claimed compensation for the cost of procuring the technical report, nor interest. If they had, CDC would surely have awarded compensation for this.

23 Time
As mentioned, the CDC handed down its decision in October 1995, one month after its initial hearing, and some three years after the damage was first noticed.

24 Role of Experts
As mentioned, the plaintiffs had a report made by another contractor, and this was used also by the CDC.

25 Outcome
CDC made the estimates in the above mentioned report the basis for its award, and the Bechs were awarded NOK 36,700.

30 Feedback
The case is available to the public.

Singapore

Chan Tan & Partners

Type A

THE PURCHASER'S OPPORTUNITY FOR RECOVERY, THROUGH LITIGATION OR OTHERWISE, FROM THE VENDOR

In Singapore, when a purchaser enters into an agreement to purchase a residential or commercial unit from a vendor who is a licensed developer, a prescribed form of agreement would be used.

Under the terms of the Sale and Purchase Agreement, the purchaser would be entitled to claim against the developer for defects in the building. The conditions of the Prescribed Form of the Sale and Purchase Agreement – for both residential and commercial developments – expressly provide that the developer 'shall erect the building unit and the common property in a good and workmanlike manner.'

The Sale and Purchase Agreement further provides that, in the event of any building defects occurring within a 12 month period from the date of delivery of vacant possession, the purchaser may deduct the rectification costs from the purchase monies held by a third party as stakeholders for this purpose.

The original purchaser may have to consider claiming against the developer in tort if the six-year limitation period for him to pursue his contractual claims against the developer has lapsed. In a claim for negligence, while the basic period is also six years from the occurrence of damage, it is possible for a claim to be brought after latent damage becomes discoverable or is actually discovered up to a maximum of 15 years from accrual of the cause of action. Purchasers would normally

Case Studies in Post-Construction Liability and Insurance, edited by Anthony Lavers. Published in 1999 by E & FN Spon, 11 New Fetter Lane, London EC4P 4EE, UK. ISBN: 0 419 24570 7

pursue their claims against developers in the courts, through litigation. There is no provision for arbitration in the Prescribed Agreement.

It is worth noting that a purchaser of a residential property from a developer may refer their dispute on building defects or non-compliance with specifications in the Agreement to the Real Estate Developers Association of Singapore (REDAS) under the REDAS Conciliation Scheme ('the Scheme'). The developer of the residential unit concerned must, however, be a member of REDAS. Under the Scheme, both the developer and the purchaser agree to refer their dispute to an independent panel called the Conciliation Panel. This Panel will hear the purchaser's complaint and the developer's response and thereafter make its recommendation or decision. The Scheme is intended to provide a simple, speedy, cost-effective and amicable method of dispute resolution. The parties may not be represented by counsel at the hearing. The developer will have to comply with the decision of the Panel or face disciplinary action by REDAS.

THE PURCHASER'S OPPORTUNITY FOR RECOVERY, THROUGH LITIGATION OR OTHERWISE, FROM THE VENDOR'S CONTRACTOR

As no privity of contract exists between the purchaser and the vendor's architect, there will be no recourse for the purchaser under contract law; instead, the purchaser would have to rely on remedies under the law of torts.

As the claim against the Contractor and the Architect involves the cost of repairs, this would amount to a claim for economic loss; the general common law position is that a claim for pure economic loss cannot be made under tort. However, the Court of Appeal, in the case of RSP *Architects & Engineers v Ocean Front Pte Ltd* ([1996] 1 SLR 113), appears to have opened the possibility that claims for economic loss in tort may be possible.

In the present case, it would appear that neither the contractor nor the architect would be held liable to the purchaser, or even to the developer. The contractor, in building according to specifications and instructions from the architect, and in the absence of defects due to poor workmanship, would have performed his obligations, under contract and in tort.

THE PURCHASER'S OPPORTUNITY FOR RECOVERY, THROUGH LITIGATION OR OTHERWISE, FROM THE VENDOR'S ARCHITECT

As the Building Research Institute report was not widely circulated until three years after the building was designed, the architect was unaware of the same and was therefore not negligent in failing to warn the developer against the use of the product. The architect was not in a position to advise as to its suitability and had voiced his reservations concerning the product. This would be a material factor in eliminating liability for the architect.

THE INVOLVEMENT (IF ANY) OF INSURERS IN THE PURCHASER'S CLAIM FOR RECOVERY, INCLUDING THE METHOD OF RESOLVING THE MATTER

Insurance policies covering defects in buildings are rarely taken out by purchasers in Singapore, and developers do not take out such policies for the benefit of the purchasers (the only construction insurance policies taken out are Contractors' All Risk Policies which cover material damage to the works and third party liability during the period of construction).

THE PROCEDURES BY WHICH THE ABOVE ARE RESOLVED, INCLUDING EXTRA-LEGAL AND INSURANCE PROCEDURES

Claims against insurers are usually resolved by litigation, unless there is an arbitration clause in the policy, in which event the dispute would be referred to arbitration.

ANY MECHANISMS EXISTING FOR UTILISING THE FEEDBACK OR EXPERIENCE FROM THE CASE TO A BODY OR BODIES WHICH MAY MAKE USE OF IT

The usual feedback would be through reported court cases. The relevant regulatory authorities, including the Building Control Department, Controller of Housing, the Singapore Institute of Architects and other professional organisations (like REDAS, if residential) would then take note of these developments.

Type B-I

The *Bayshore Park* case

1 The Plaintiff
The plaintiff was the Management Corporation of the Bayshore Park Condominium, a body established to manage the condominium and its common property.

2 The Works
The high-rise residential property was a framed construction, with no innovative techniques and normal exposure to the elements. The value was in the region of SGD 100 million.

3 Procurement Method
The plaintiff contracted with a developer, Ocean Front Pte. Ltd. under traditional procurement arrangements, and allocated design and supervision tasks to a firm of consultant engineers and architects.

4 Insurance
Only a Contractors' All Risk insurance was taken out.

5 Damage or Loss
The plaintiff noticed spalling of concrete in the ceilings of car parks and water ponding in some lobbies. The loss, due to poor workmanship, was economic.

10 Process after Discovery
The plaintiff initiated a claim against the developer who joined the architects, the building contractors and the structural engineers as third parties.

The defendants applied to the High Court for determination of a preliminary issue, namely whether the plaintiff was barred from claiming the costs of repairs, as the claim was based on economic loss. The defendants appealed the verdict from the High Court.

11 Technical reports
Not known.

20 Resolution
The High Court held that the plaintiff could claim in tort for the costs of repair. The Court of Appeal affirmed the decision of the lower court.

21 Costs, 22 Information, 23 Time, 24 Role of experts: Not known.

25 Outcome
See 20.

Type B–II

The *ABC Villas* case

1 The Plaintiff
The plaintiff was a contractor.

2 The works
Eight bungalows were built as framed construction with pitched roofs, with no innovative techniques and normal exposure to the elements.

As the bungalows were located on a sloping site between two roads, the project included terracing of the site and incorporating retaining walls supported by large diameter bored piles at the major steps in level; the project included monitoring the piles for lateral movement during the construction period.

The contract value was in the region of SGD 16 million.

3 Procurement method
Not relevant.

4 Insurance
Only a Contractors' All Risk insurance (CAR) was taken out.

5 Damage or Loss
The damage occurred before completion and consisted in lateral movement in excess of that permitted by the original design.

During the construction period, the water mains of the higher of the two roads was damaged and created a leakage. Although informed of this event, the local authorities did not effect immediate repairs. The ground above and behind the bored pile wall then rose a further 3 metres above the level prescribed.

10 Process after Discovery

When the main contractors noted the movement in the bored pile wall beyond the originally stated maximum permissible, they instigated a second site investigation. This revealed that the soil near the bored piles was weaker than the original investigation had shown. The deflection was attributed to the improperly compacted fill or poor soil the wall, or both.

The contractor was instructed to carry out rectifying work, and did this at a cost of some SGD 1 million. The contractor claimed for this under its CAR.

The insurers conducted their own investigation and declined the contractor's claim.

11 Technical reports

See above.

20 Resolution

If the insurers settle the main contractor's claim, they may, through subrogation, claim against the developers.

21 Costs, 22-Information, 24 Role of experts and 25 Outcome: Not known.

Spain

Manuel Olaya Adán

Type A

THE PURCHASER'S OPPORTUNITY FOR RECOVERY THROUGH LITIGATION OR OTHERWISE, FROM THE VENDOR

A contract between a purchaser and a vendor implies application of the Civil Code (CC) Arts 1484–1485, and of the Code of Trade, Arts 336–337.

In the case under scrutiny, the Code of Trade would offer no mechanism for recovery in court.

According to the CC, however, the vendor must repair hidden defects if the purchaser has no relevant experience, also if the vendor did not know about the defect at the moment of sale, and if the defect is 'hidden', *vicio oculto*, but not for obvious or patent defects.

Arbitration would be relevant only if it was explicitly stated to be so in the contract.

THE PURCHASER'S OPPORTUNITY FOR RECOVERY, THROUGH LITIGATION OR OTHERWISE, FROM THE VENDOR'S CONTRACTOR OR ARCHITECT

CC Art 1591 (about Decennial Liability in construction) is not applicable because the building is not in a state of *ruina*. The concept of *ruina* is akin to 'the building not functioning', and would not cover cosmetic flaws. Nor would the Law 22/1994 concerning liability for defective products (de-

Case Studies in Post-Construction Liability and Insurance, edited by Anthony Lavers. Published in 1999 by E & FN Spon, 11 New Fetter Lane, London EC4P 4EE, UK. ISBN: 0 419 24570 7

rived from the European Directive 1985/374) be applicable, since no health or safety matter is at issue.

THE INVOLVEMENT (IF ANY) OF INSURERS IN THE PURCHASER'S CLAIM FOR RECOVERY, INCLUDING THE METHOD OF RESOLVING THE MATTER

If the vendor carried an insurance policy, the purchaser's claim for recovery could involve the vendor's insurance company. Although neither the building's structural soundness nor any other crucial aspect of its performance was affected, negotiation with the vendor's insurance company could prove the most effective way to recovery.

THE PROCEDURES BY WHICH THE ABOVE ARE RESOLVED, INCLUDING EXTRA-LEGAL, LEGAL AND INSURANCE PROCEDURES

In this case, there are two contracts with legal relevance: the contract for acquisition of the glass-fibre reinforced concrete product for the cladding, and the contract relating to the purchase of the building.

Both contracts could contain arbitration clauses. If so, an extra-legal solution could be attempted, involving the supplier of the product, the developer and the purchaser.

In legal proceedings, the purchaser's claim would, as mentioned, be founded on CC Art s 1484 and 1485. The same articles could give the developer the right to claim against the supplier of the cladding, possibly involving also the supplier's insurer, if any. No extra-contractual liabilities would be called upon in this case.

ANY MECHANISM EXISTING FOR UTILISING THE FEEDBACK OR EXPERIENCE FROM THE CASE TO A BODY OR BODIES WHICH MAY MAKE USE OF IT

Court proceedings, but not extra-legal solutions such as arbitration, are available to the public.

Type B–I

The *Leaking Roof* case

1 The Plaintiff
The plaintiff was an architect who bought a five year old single family house, designed and built by someone else, for his residence.

2 The works
The plaintiff intended to and did use the property for his own residence.

3 Procurement Method
The plaintiff bought the property, which had been completed in 1984, five years after completion.

4 Insurance
There was no property insurance.

Both the architect who had designed the house and the *arquitecto técnico* in charge of supervision of the works carried professional liability insurance.

5 Damage or Loss
The plaintiff, while living in the house, noticed that the roof leaked in some places, and claimed compensation for the cost of reparation and economic loss.

The damage had developed progressively. The underlying defect was poor quality of the mortar between roofing tiles. The plaintiff held that this was due to poor supervision (of which the *arquitecto técnico* was in charge) during the works.

10 Process after Discovery
The owner initiated the claim against both the contractor and the *arquitecto técnico*.

11 Technical Report
The plaintiff, as an architect, produced his own report which, together with the claim, he lodged as a notary document.

The defendants did not produce or commission technical reports.

20 Resolution
Initially, the plaintiff intended to come to an agreement with the contractor, but having received no reply from this party for two months, the plaintiff then initiated legal action against both the contractor and the *arquitecto técnico*.

The contractor did not use legal counsel; the *arquitecto técnico*'s liability insurer provided legal counsel.

21 Costs
The plaintiff's legal costs were partly paid by the *arquitecto técnico*'s insurer, cf. 23 (below).

22 Information
Because of the progressive process of the damage, it was not possible to create a detailed pathology.

23 Time
The court accorded the two defendants four months to reply to the plaintiff's claim. After this period had expired, the case went on for two years, and was concluded by an extra-legal agreement between the parties.

24 Role of Experts
All parties provided their own expertise.

25 Outcome
See 23.

Type B–II

The *Parking Garage* case

1 The Plaintiff
The plaintiff was an association of private persons, an Owners' Association.

2 The works

The parking garage, built in 1983, was situated under a public urban space.

3 Procurement Method

The developer cum builder both designed and built the garage.

The private persons bought the property from him and later created their Owners' Association, a legal entity.

4 Insurance

There was no property insurance.

5 Damage or Loss

The plaintiff discovered soon after the sale that water penetrated through the vault in several places. After some two years, the damage was extensive.

10 Process after Discovery

The Owners' Association initiated an action based on article 1591 of the Civil Code (decennial liability in construction) against the builder–developer.

11 Technical Reports

The Owners' Association commissioned a number of technical reports, which were transmitted to its legal counsel.

20 Resolution

The Owners' Association won the case in the first instance, and the defendant appealed.

21 Costs

Not known.

23 Time

The case took two years in the first instance and a further year in the appeals court.

25 Outcome

The appeal also went in favour of the Owners' Association, but as the defendant was bankrupt, the plaintiff never received compensation.

Sweden

Type A

Anders Kleberg

THE PURCHASER'S OPPORTUNITY FOR RECOVERY THROUGH LITIGATION OR OTHERWISE, FROM THE VENDOR

According to Swedish law, the vendor must inform the purchaser of 'hidden defects' (i.e. defects that cannot be noticed during a thorough examination) known to the vendor. On the other hand, the purchaser must conduct a thorough examination, and cannot turn against the vendor concerning defects which could have been discovered during this examination.

This being so, if the cracks could have been discovered during an examination, then the purchaser would have only a limited chance for recovery. If, on the other hand, the cracks could not have been noticed at the time of the transfer, some remedies for damages from the vendor would exist.

The purchaser did not fulfil its duty to inspect the building; and the vendor acted in good faith.

THE PURCHASER'S OPPORTUNITY FOR RECOVERY THROUGH LITIGATION OR OTHERWISE, FROM THE VENDOR'S CONTRACTOR OR ARCHITECT

Without a contractual relationship between the purchaser and the vendor's contractor or the vendor's architect, Swedish law grants no recovery for pure economic loss. This principle of privity of contract has, however,

Case Studies in Post-Construction Liability and Insurance, edited by Anthony Lavers. Published in 1999 by E & FN Spon, 11 New Fetter Lane, London EC4P 4EE, UK. ISBN: 0 419 24570 7

been subject to some changes: When the situation is akin to a contractual one, a quasi-contractual relation has occasionally been construed to exist.

THE INVOLVEMENT (IF ANY) OF INSURERS IN THE PURCHASER'S CLAIM FOR RECOVERY, INCLUDING THE METHOD OF RESOLVING THE MATTER

Generally speaking, an insurance company would not cover this kind of defect, since it was built in and therefore existed before the purchase.

The property policy known as Insurance Against Latent Defects usually runs for ten years from the time of hand-over. The purpose of this insurance is to protect a party to the contract from hidden defects, but cosmetic flaws fall outside its scope.

THE PROCEDURES BY WHICH THE ABOVE ARE RESOLVED, INCLUDING EXTRA-LEGAL, LEGAL AND INSURANCE PROCEDURES

This issue would fit within the usual court system unless there was an arbitration clause or both parties wanted the case to be settled by arbitration. The case could also be dealt with, by alternative dispute resolution, or settlement out of court. It is worth noting that a special tribunal to settle professional liability cases has been set up by the Organisation of Architects and Consulting Engineers.

ANY MECHANISMS EXISTING FOR UTILISING THE FEEDBACK OR EXPERIENCE FROM THE CASE TO A BODY OR BODIES WHICH MAY MAKE USE OF IT

None.

Type B

Anders Kleberg with Kåre Eriksson

Type B–I

The *Heating* case

1 The Plaintiff
The plaintiff was a consumer and client organisation, one of the country's largest developers of bungalows and flats, with its own technical expertise. It belonged to the private sector.

2 The Works
The project was an apartment building.

3 Procurement Methods
The client appointed the main contractor and the main consultant for all the design work.

The main contractor appointed a sub-contractor for the heating system.

The main consultant appointed a sub-consultant for design of the heating system.

4 Insurance
The designer carried professional indemnity insurance, covering a designer's liability according to General Conditions for Architects and Consulting Engineers *ABK 87*.

The main contractor also carried a common professional indemnity insurance.

5 Damage or Loss
Defects in the heating system were noticed by the tenants and reported to the client. It was found that the system did not work properly, though the cause for this was unknown.

Correcting the defect was estimated to cost around SEK 1.5 million.

10 Process after Discovery
The Client blamed the main consultant for faulty design and notified its insurer. The main consultant turned against the sub-consultant who notified its insurer.

11 Technical reports
The client commissioned and paid for a technical report. The client's insurer commissioned and paid for a technical report from a different expert.

20 Resolution
A negotiated settlement was reached.

The negotiations on behalf of the main consultant were handled by the firm's staff and a lawyer from its insurer, with some assistance from an external lawyer. The client appointed its own lawyer. The sub-consultant's insurer was represented by a specialist in construction and an architect.

21 Costs
Each party met its own costs.

22 Information
All parties had access to all information. At the time of settlement it was, however, not clear what had caused the problem.

23 Time
Thirty hours of investigation and discussion followed by five hours to reach the final settlement.

24 Role of experts
The parties used their in-house experts to a certain extent but relied mainly on external experts.

25 Outcome
The main consultant's insurer paid around SEK 600,000 to the client, and received around SEK 300,000 from the sub-consultant's insurer and around SEK 80,000 from the main consultant, equivalent to the excess in its liability insurance.

30 Feedback
No part of the case is available to the public.

Type B–II

The *June-Expressen* case

1 The Plaintiff
The plaintiff was a company which occasionally commissioned construction works. It belonged to the private sector.

2 The Works
An industrial plant was built on a site about which it was known that the soil might contain soft strata. The design for the foundation was based on a soil examination carried out for an adjacent plot. On the site, three bore holes were made to ascertain that the soil conditions were the same as on the adjacent plot.

3 Procurement Methods
Not relevant.

4 Insurance
Not relevant.

5 Damage
After completion, a wall subsided. The reason for this was erroneous design of the foundation due to the incomplete soil investigation.

10 Process after discovery
The client sued a sub-consultant for the cost of strengthening a wall.

11 Technical Reports
The plaintiff had a technical report made, at his own expense. The defendant did not commission a report.

20 Resolution
The plaintiff lost the case in court and appealed.

21 Costs
Not known. Normally the losing party meets both parties' legal costs.

22 Information
Not known.

23 Time
Not known.

24 Role of experts
An expert was, as mentioned, appointed by the plaintiff.

25 Outcome
The appeal court did not change the lower court's findings.

Switzerland

Type A

Viktor Aepli and Roland Hürlimann

THE PURCHASER'S OPPORTUNITY FOR RECOVERY THROUGH LITIGATION OR OTHERWISE, FROM THE VENDOR

1 Law Sources (Law of sale and purchase)

1.1 The purchase is a purchase of real estate under Ss.216ff. of the Swiss Code of Obligations (SCO). As regards warranty, those statutory provisions contain but few rules. On that point, SCO S.221 makes reference chiefly to the provisions concerning the sale and purchase of movable assets.

1.2 The so-called 'object warranty' is regulated by SCO Ss.197ff. By SCO S.197 para.1, the vendor is liable to the purchaser for the specified properties as well as for the absence of any physical or legal defects which cancel or reduce the value of the object or the fitness thereof for the implied use. By para.2 of the same section, the vendor will still be liable if he was unaware of the defects.

1.3 By SCO S.199, the warranty may be contractually excluded; yet, such a covenant will not be effective if the vendor has deceitfully failed to inform the purchaser of the warranty defects.

1.4 By SCO S.200, para.2, the vendor will not be liable for defects of which the purchaser was aware at purchase. For defects which the purchaser ought to have discovered in applying ordinary diligence, the vendor will not be liable unless he promised that the building would be free from them (SCO S.200 para.2).

Case Studies in Post-Construction Liability and Insurance, edited by Anthony Lavers. Published in 1999 by E & FN Spon, 11 New Fetter Lane, London EC4P 4EE, UK. ISBN: 0 419 24570 7

1.5 The purchaser is required to inspect the qualities of the purchased object and, if he finds defects for which the vendor is liable under the warranty, to notify him immediately (SCO S.201, para.1). If the purchaser fails to make such inspection or notification, the purchased object will be presumed to have been accepted, except for defects which were not detectable by ordinary inspection. If, subsequently, such defects are found, notification must be made immediately on discovery, failing which the object will be presumed to be accepted in respect of those defects also (SCO S.201, para. 3).

1.6 After giving notice of the existence of the defect, the purchaser will have the option either to cancel the purchase by notice of rescission or to claim indemnity for the diminution in value by notice of diminution (SCO S.205, para.1). Nevertheless, the court will be free merely to award diminution indemnity if the circumstances do not justify cancelling the purchase.

1.7 On rescission of the sale-purchase contract, the purchaser must return the object, and the vendor must refund the paid price plus interest. If the vendor has acted culpably, he will be bound also to give indemnity for further loss (SCO S.208). Such additional indemnification also applies in a case of diminution (SCO S.97).

1.8 By SCO S.219, para.3, the warranty obligation for defects in a building lapses with the expiry of five years from acquisition of ownership.

2 Substantive Issues

2.1 A first point to examine is whether, under SCO S.197, para.1, the building contains a defect at all. True, the developer did not expressly warrant to the Bank that the cladding tiles were free of defects, and so it did not warrant any particular quality. Yet, a defect may be present if an implied property is lacking in the building. The developer will be liable for the absence of such properties if the economic value of the building is thereby materially diminished. In the present case, the defective tiling has considerably impaired the outward appearance of the building, and so diminished its value. We will therefore assume that the building presents a defect under SCO S.197.

2.2 There is nothing in the documents to suggest that the parties ruled out or restricted the developer's warranty obligation. That is not the usual practice, but in what follows, we will assume such a covenant to be absent.

2.3 The Bank, at the time of purchase, was not aware of the defects in the building. The question now is whether that Bank, in applying ordinary diligence, ought to have discovered them; the documents show that an experienced surveyor would have detected the cracks and assessed the potential deterioration. Yet, the Bank was not bound to inspect the building before signing the sale-purchase contract. So the Bank was neither aware of the defects nor required to be aware of them.

2.4 A successful claim under the warranty is conditional on timely notice of the defect. For the purchase, the Bank would first have had to inspect the building in time or, if it lacked the necessary expert knowledge, commission a surveyor accordingly. In the present case, the first cracks had appeared before signature of the contract and before transfer of ownership; yet the Bank was not required to search for latent defects, and so it was not bound to have the building inspected by a surveyor. Moreover, at the time of purchase, information on the problem inherent in the particular tiling was not readily available. Generally, the Bank was basically not required to enlist a surveyor where there was no specific reason to suspect improper performance. We will therefore assume that inspection of the building was made within time and with proper diligence, and that the Bank did not accept the building with respect to those defects.

2.5 With the hairline cracks spreading, the appearance of the building deteriorated after the purchase. It was only then that the Bank became aware of the defect. It was to give notice of the defect immediately on discovery. Such a latent defect is not deemed discovered until detected beyond all doubt. In the present case, though, it is a moot point at which time the defect became discoverable by the Bank, as the appearance of the building deteriorated slowly. We will assume that the purchaser was to give notice of the defect when it decided that it could not leave the building in such a state. If then the purchaser failed to notify the developer (vendor) immediately (i.e. within one to four days), it would lose its rights, but, as said, there is likely to be some latitude in the present case as to the time of discovery of the defect.

Statute law does not specify any particular form for the notice of defect, yet the rule is to notify the vendor in writing, if only to provide evidence for later use. The content of the notice must include a specification of the defects found, to enable the developer to inspect the building or have it inspected. General comments merely expressing the Bank's dissatisfaction with the building would not suffice.

2.6 There are now two alternatives open to the Bank: either to demand rescission of the contract, or to claim indemnity for the diminution in value. The Bank is not required to decide for one or the other immediately, but is basically free to exercise the option any time during the limitation period (five years for buildings).

In some cases, the choice may be refused. The court may, if the circumstances so justify, merely award indemnity (damages) for the diminution in value.

Basically, though, the court will not decide in favour of indemnity instead of rescission unless the Bank can reasonably be expected in good faith, by market standards, to retain the building despite the defects, and in particular unless rescission would set the developer at a disadvantage disproportionately greater than the advantage to the Bank. In the present case, we will assume that the Bank can sue both for rescission and (alternatively) for diminution indemnity.

It is to be noted that the law of sale-purchase (unlike the law governing work contracts) does not grant the developer any opportunity for remedial work, or any right to the Bank to demand such work.

2.7 The developer must submit to rescission or diminution indemnity, accordingly, even if no culpability is involved. It is therefore immaterial whether the developer knew of the defects or whether, in the nexus between developer and contractor–Architect, one of the latter might be answerable for the defects.

2.8 If the developer is found culpable, he will be ordered to indemnify the Bank for additional loss. Such loss would include loss of rent, for instance; but it would not include the cost of repair work which the Bank might commission of its own accord, because the loss in value due to the defects is to be recovered by the action in diminution.

2.9 In summary: The Bank can claim against the developer as vendor. The Bank can sue for rescission or for diminution damages. The Bank cannot demand remedial work, nor have such work done at the developer's cost. The Bank's right to rescission or diminution indemnity subsists even without any culpability on the developer's part. Special points to be noted are the short time limit for notice of defect and the five-year limitation period.

THE PURCHASER'S OPPORTUNITY FOR RECOVERY, THROUGH LITIGATION OR OTHERWISE, FROM THE VENDOR'S CONTRACTOR

1 Preliminary points

There is no contractual nexus between the contractor and the Architect on the one hand and the Bank on the other. Any liability of the Architect or contractor for defects in the building cannot therefore be established through contract law. It would be different if the developer, under the sale-purchase contract, had assigned to the Bank its claims under its work contracts with the Architect and contractor. Such assignment can still be made after signature of a sale-purchase contract, and the Bank could demand assignment of those claims, especially as the law governing work contracts offers some advantages over the law of sale-purchase.

Irrespective of any assignment of claims for defects to the purchaser, the latter may have claims against the contractor in tort (SCO Ss.41ff.). A discussion of that aspect is beyond the scope of this Case Study. Suffice it to say that liability under SCO Ss.41ff. would require tortious conduct and culpability on the contractor's part, and the Bank would have to prove it; and though there might be a case for culpability, the Bank would be taking a considerable risk of losing the case.

If the developer has assigned its rights to the Bank, the first point to examine is the nexus between contractor and developer as building owner. That nexus may be of two kinds, viz. subject to the statutory rules (work-contract law, SCO Ss.363ff.) or subject to the SIA (Swiss Engineers' and Architects' Association) Standard 118. The SIA Standard 118 is a specimen (i.e. standard form) contract which is usually adopted for construction work. We will now outline, first, the statutory (optional) rules and, secondly, the position under the SIA Standard 118.

1.1 contractor's liability under statutory rules

1.1.1 Law sources (work-contract law):

Under a work contract, the contractor is carrying out works (here a building), and the client (building owner) undertakes to pay the contractor for it. The contractor's liability is governed by SCO Ss.367ff.

1.1.2 By SCO S.367, the client is required, as soon as convenient in the ordinary course of business, to inspect the qualities of the product and notify the contractor of any defects found (Hürlimann, *Architekt als Experte*, p.436).

If the client expressly or tacitly accepts the product, the contractor will be released from liability, except for defects which were not detectable at acceptance or which were wilfully concealed by the contractor.

If the client omits the statutory test or fails to notify, tacit acceptance will be presumed.

1.1.3 If the defects do not show until later, notice must be given immediately on discovery, failing which the product will be presumed to be accepted in respect of those defects also (SCO S.370, para.3).

1.1.4 The client's rights concerning defects found will lapse if the client himself is to blame for the defects through having given instructions for execution against the contractor's express warning (SCO S.369).

1.1.5 If notice of defect has been given within time, and if the client himself is not to blame for the defects, three options will be open to the client under SCO S.368.

5.1 He may refuse acceptance if he cannot reasonably be expected to accept (rescission).

5.2 He may deduct from the contract price an amount corresponding to the diminution in the product value (diminution).

5.3 He may demand remedial work free of charge (remedial work).

In all three cases, the contractor, if culpable, may also be sued for damages.

On work contracts for the construction of buildings, the right to rescission is restricted. If removal of the building involves unreasonable disadvantages, the client will merely have the choice between diminution indemnity and remedial work (SCO S.368, para.3.).

1.1.6 The building owner's claim for defects in the building will lapse with expiry of five years from acceptance (SCO S.371, para.2.).

2 Substantive issues

2.1 A first point to be examined is whether the product contains a defect under SCO Ss.367ff. A product defect establishing the contractor's liability is a variation of the product from the contract. The defect may be the absence of a specified or implied property (Gauch, *Werkvertrag*, nn.1355ff.). Implied properties are such as relate to 'value quality' and fitness for use. As regards 'value quality', the contractor, unless otherwise agreed, is to produce a product having the standard nature (Gauch, *Werkvertrag*, n.1409). He is therefore bound to use 'thoroughly good materials'. By unsound tiling, the building has suffered a loss in value, and is therefore to be rated defective.

2.2 In the present case, the defect was not detectable at acceptance of the building. It is therefore a latent defect under SCO S.370, para.3. The developer, or the Bank, to which the developer's rights concerning defects found were assigned, was to notify the contractor of the defect immediately on discovery. On that point, Swiss Federal court precedent tends to be strict (Gauch, *Werkvertrag*, n.2180). Latent defects in a building are presumed to be discovered as soon as the client is certain of their presence and has established their existence beyond a doubt; that may mean consulting an expert. Latent defects which show up only gradually are not presumed discovered at the appearance of the first signs: rather, they are so presumed once the client has realised the significance and implications of the defects (Gauch, *Werkvertrag*, n.2182).

If the client fails to give notice of a latent but discovered defect, the product will be presumed accepted in respect of that defect (Gauch, *Werkvertrag*, n.2185). The Bank therefore would have to act immediately and notify the contractor of the defect found.

2.3 In the present case, however, the question arises whether the developer did not lose its rights concerning this defect by ordering the use of unsound tiling and, against the Architect's warning, insisting on using it (SCO S.369) (see also below). Another point is whether the fact that the Architect is held to be an assistant of the developer does not result in ruling out the contractor's liability or in the contractor's partial release.

2.3.1 The developer insisted on using the tiling. It thereby ordered the contractor to use the particular material (cf. Gauch, *Werkvertrag*, nn.2015ff.). We will assume that the tiling thus specified was by its nature unfit for a building free of defects. That is not altered by the fact that, at the time of construction, neither the contractor nor the developer was aware of such unfitness. Therefore, the contractor will not be liable if it warned the developer and if the latter insisted on using the tiling (Gauch, *Werkvertrag*, n.2018).

There is nothing in the documents to suggest that the contractor gave warning; rather, it was the Architect who voiced doubts. By SCO S.369, however, the client's rights concerning defects found will be ruled out by the client's own culpability, and so the contractor could also argue the warning given by the Architect. It remains to be examined whether the Architect gave proper warning (in the contractor's favour) according to SCO S.369.

A warning must be given expressly. It must point out that the instruction is misconceived because its execution may result in a defect in the building (Gauch, *Werkvertrag*, n.1940). Such a warning was given neither by the Architect nor by the contractor. True, the Architect declared that he had no experience of the product in question. Yet, that alone is not proper warning, as no reference was made to the chance of a defect arising or to any facts prompting the Architect's misgivings. The fact that the Architect was unaware of the paper cited is therefore to the disadvantage of both the Architect and the contractor. Those two were the experts concerned, and the fact that they were ignorant of the particular building material cannot be held against the developer. It might be different if the Architect had not simply said that he had no experience of such materials, but had expressly declined liability.

We will therefore assume that the developer's insistence on using the unsound tiling did not destroy its rights concerning defects found.

2.3.2 In the present case, the Architect is an assistant of the developer. The developer thus led the contractor to believe that it could rely on the specialist knowledge of the developer's assistant (Gauch, *Werkvertrag*, n.1293). The Architect's knowledge is therefore to be reckoned as the developer's own. By SCO S.369, however, sufficient culpability will be present only if the developer alone is answerable for the defect (Gauch, *Werkvertrag*, nn.1916ff.). Thus such culpability will not be present if the contractor contributed to the generation of the defect by a culpable breach of its duty of diligence (Gauch, *Werkvertrag* n.1919).

The contractor's duty of diligence includes the duty to warn (Gauch, *Werkvertrag*, nn.833ff.). The contractor failed to discharge it and so supplied an important factor for the defect to arise. Not having given warning, the contractor could only escape liability if it had not known of the unsuitability of the instruction and was not required to know (Gauch, *Werkvertrag*, n.1963 and nn.1967ff.). At the time of construction, however, a relevant paper from the building research institute existed; with reasonable effort, the contractor could have obtained the information on the unsoundness of the tiling desired. As it did not do so, the developer (or its assistant) is not alone to blame under SCO S.369. Hence the developer has not lost its rights concerning defects found, even if the Architect's knowledge is treated as its own.

2.3.3 Although, in the present case, we must reject sole culpability on the developer's part (which would rule out the contractor's liability), co-culpability on the developer's part may result in reducing the contractor's liability (cf. Gauch, *Werkvertrag*, nn.2050ff.). Such co-culpability is present here because the developer gave the instruction from specialist knowledge (i.e. that of its Architect).

As regards loss consequent on defects, the court may take into account any co-culpability of the developer under SCO S.44, para.1. For the right to diminution and remedial work, SCO S.44, para.1 is to be applied accordingly (Gauch, *Werkvertrag*, nn.2061ff.): Diminution will not be allowed in full: the contractor has not to bear alone the cost of remedial work, but may claim part from the developer. As for rescission, the developer's co-culpability implies that it can be more reasonably expected to accept the building. If rescission is nevertheless allowed, the contractor will receive part of the contract price. In the present case, however, the developer will anyway be deprived of the right to rescission; and so the Bank will have to accept a reduction of its claims against the contractor.

2.4 The Bank can therefore choose between diminution indemnity and remedial work. It is doubtless not entitled to rescission under SCO S.368, para.3, given the fact that a building has been constructed and that only its outward appearance looks somewhat shabby.

2.5 If the Bank raises its right to remedial work, the contractor will be bound to do the remedial work. Yet the contractor itself is not entitled to insist on doing the remedial work: the choice between remedial work and diminution indemnity (and rescission) is a matter for the Bank alone (Gauch, *Werkvertrag*, n.1711). The contractor is basically required to do the remedial work free of charge, yet the Bank will have to accept a reduction of its claims (above). The right to remedial work is subject to the condition that removal of the defects is practicable at all. That is likely to be so here: the defective tiling can be replaced, though at some cost. Moreover, by SCO S.368, para.2, a further condition is that the remedial work should not involve excessive cost to the contractor, (Gauch, *Werkvertrag*, n.1749). The cost will be deemed excessive if it is disproportionate to the benefit to be gained by the Bank by removal of the defect (Gauch, *Werkvertrag*, nn.1748ff.). Here, it is difficult to assess whether the Bank can still demand remedial work. True, the Bank can also argue prestige appeal, yet where it is a matter of purely visual flaws, the costs of remedial work is likely to be considered excessive, unlike the case where fitness for use is in question (Gauch, *Werkvertrag*, n.1757).

2.6 The Bank may instead accept the building as it stands and deduct from the contract price an amount corresponding to the diminution in value. As the developer has already paid the contractor, and as the rights concerning defects found are raised by the Bank, the latter's claim will be for refund of the excess paid (Gauch, *Werkvertrag*, n.1617).

A claim in diminution is conditional on the building having incurred a diminution in value. Such a diminution in value is determined on objective criteria. Here, it is likely that there will be found to have been diminution in value. A further condition is that the defective building should not be entirely valueless (Gauch, *Werkvertrag*, no. 1639). That condition is also present here, and so the Bank can claim in diminution.

2.7 As said, the Bank's rights concerning defects found subsist independently of any culpability on the contractor's part. In that respect, it is therefore immaterial whether or not the contractor knew of the unsoundness of the tiling. Yet, the contractor's culpability will be relevant to its liability for loss consequent on defects. Consequential loss may consist in loss of rent, for instance. By SCO S.368, para.1, the contractor will be liable for such loss if found culpable; negligence is sufficient. The Bank need not prove the contractor's culpability: it is for the contractor to prove the contrary (SCO S.97, para.1). Therefore, the point to be decided is whether the contractor ought to have known of the paper cautioning against the use of such tiling, and whether non-knowledge thereof amounts to culpability. The point cannot definitely be settled on the documents available.

The diligence required of the contractor is also determined by the accepted state of the art prevailing at performance of contract (Gauch, *Werkvertrag*, n.1888). The paper had already been made available at the time of construction. The contractor knew that such materials were unfamiliar to it. It could reasonably have been expected to obtain relevant information. So, although the contractor did not go against the accepted state of the art when constructing the building, it did fail to apply the necessary diligence and so it incurred culpability.

2.8 By SCO S.371, para.2, the Bank's claims lapse with the expiry of five years from acceptance of the building (Gauch, *Werkvertrag*, nn.2195ff.). In the present case, that period has not yet expired, and so it is basically still open to the Bank to sue the contractor (provided that the assignment is valid).

3 contractor's liability under SIA Standard 118

3.1 Law Sources (SIA Standard 118)
The SIA Standard 118 is a private law standard form or specimen contract published by the Swiss Engineers' and Architects' Association, SIA (Gauch, *Kommentar, Einleitung*, n.2)). It contains general conditions for construction work, which by their nature are specimen terms. They do not have the general validity of statute law, but are merely valid (if at all) between parties which have selected them (Gauch, *Kommentar, Einleitung*, n.11). Given the wide acceptance of the SIA Standard 118 in the Swiss construction industry, it is appropriate to discuss its rules on liability.

3.2 By Art. 166 of the SIA Standard 118, 'defect' means that the building lacks a promised or otherwise agreed quality, or that the building lacks a quality which the client was entitled, in good faith and without special agreement, to expect. Such *bona fide* implied qualities also relate to 'value quality': the contractor owes a building of a 'value quality' corresponding to *bona fide* standards (Gauch, *Kommentar*, n.7 on Art. 166). However, the term 'defect' does not cover a non-contractual state of the building for which the client or any of its assistants is entirely to blame. The client will not be culpable if the contractor has failed in its general duty to notify or to warn (Art. 166, para.4).

3.3 The contractor is required to construct a building free from defects (Art. 165, para.1). Its liability in that respect subsists regardless of any culpability on the contractor's part. Exceptions are cases of client's culpability (Art. 166, para. 4, though in such cases, by the terminology of the said Standard, there would be no 'defect'), and of loss consequent on defects (Art. 171).

3.4 On discovering a defect, the client has at first only the right to request the contractor to remove the defect within a reasonable time (Art. 169). In contrast with the law governing work contracts, the contractor has the right (and the obligation) to remove the defect.

3.5 If the defects are not removed within the time specified by the client, the latter may choose one of the following options:

.5.1 He may further insist on remedial work, provided that the cost of such work, related to its interest in removal of the defects, is not excessive (Art. 169, para.1, head 1).

.5.2 He may deduct from the contract price an amount corresponding to the diminution in the value of the building (diminution, Art. 169, para.1, head 2).

.5.3 He may rescind the contract. Yet, this option will only apply if removal of the building does not involve disproportionate disadvantages to the contractor, and if the client cannot reasonably be expected to accept the building (Art. 169, para.1, head 3).

3.6 If the defect has led to consequential loss, the client will additionally have a right to indemnity. The contractor will be released from such indemnification if he proves that he is not to blame (Art. 171).

3.7 The SIA Standard 118 distinguishes between two time limits for notice of defect:

.7.1 During the first two years from hand-over of the building, the client may give notice at any time of defects discovered and becoming apparent during that period (Warranty Period, Art. 173, para.1). With expiry of the Warranty Period, the client's right to give notice of previously discovered defects lapses (Art. 178).

.7.2 Within three years of expiry of the Warranty Period, the client may give notice of defects discovered after the Warranty Period ('latent defects', Art. 179). Yet, he must notify the contractor of such defects immediately on discovery.

.7.3 The client's rights concerning defects found lapse five years after acceptance of the building (Art180: Statute of limitation period).

4 Substantial issues

4.1 In the present case, the building has a defect under SIA Standard 118, Art. 166, as the value of the building is considerably diminished.

The developer insisted on using the unsound tiling. A question here is whether that fact does not override the unsoundness of the tiling used (Art. 166, para.4). By that article, the developer will not be culpable if the contractor fails to notify it or give warning. The point is regulated by Art. 25. It is likely – as in the assessment under the law governing work contracts – that no proper warning was given. If so, the presence of a defect under Art. 166 is to be affirmed.

4.2 The building was sold about two years after completion. The Bank did not discover the defects until after purchase. We will therefore assume that the Warranty Period under Art. 173 had already expired by the time of the discovery of the defects. The Bank was therefore to give notice of the defects immediately on discovery. The documents available do not reveal whether or not such notice was given. We will assume that notice was given within time.

4.3 The Bank was now required to set the contractor a reasonable time limit by which to remove the defects. If the contractor fails to comply, the Bank can basically choose between remedial work, diminution indemnity and rescission.

4.4 As regards remedial work, reference is here made to the comments on work contracts. In the case as put, it is not possible to decide whether such remedial work would involve excessive cost to the contractor (Art. 169, para.1, head 1).

4.5 The Bank cannot exercise its basic right to rescission because removal of the building work would involve a disproportionate disadvantage to the contractor, and because the client may reasonably be expected to accept the building (Art. 169 para.1, head 3).

4.6 It now remains open to the Bank to argue diminution and to claim refund of part of the contract price (Art. 169, para.1 head 2). On that point also, reference is here made to the comments on work contracts.

4.7 By the SIA Standard 118, the Bank's rights concerning defects found likewise subsist independently of any culpability on the contractor's part. There again, it is immaterial whether or not the contractor was aware of the unsoundness of the tiling. As under work-contract law, however, indemnity for loss consequent on defects implies culpability on the contractor's part. An indemnity obligation of the contractor for consequential loss is likewise to be affirmed under the SIA Standard.

4.8 By Art. 180, the Bank's claims will lapse with the expiry of five years from hand-over of the building. In the present case, that period has not yet expired.

THE PURCHASER'S OPPORTUNITY FOR RECOVERY, THROUGH LITIGATION OR OTHERWISE, FROM THE VENDOR'S ARCHITECT

For the nexus between Bank and Architect, the same is to be said as for the nexus between Bank and contractor: there is no contract between those parties. The Bank can therefore not claim against the Architect unless the developer assigned its claims to the Bank. Basically excepted are claims in tort (SCO Ss.41ff.); such claims are not under discussion here.

If the developer's rights against the Architect have been assigned to the Bank, the first point to decide is what legal nexus existed between developer and Architect.

The documents available do not reveal the terms of contract between developer and Architect. The contract with the Architect depends entirely on what the parties actually agreed (Gauch, *Architekturvertrag*, n.30). The Architect's contract may be a simple agency contract, or a work contract, or a type of contract not regulated by statute law. The so-called 'blanket contract' – by which the Architect undertakes all architectural services required for the construction of a project – has been defined by the Swiss Federal Court as a mixed contract consisting partly of a simple agency contract and partly of a work contract (Gauch, *Architekturvertrag*, n.38). If, under a blanket contract, only some specific services of the Architect are in dispute, that definition may result in a division of the legal consequences; thus, for instance, the liability for a design error would have to be assessed under work-contract law, while that for careless construction supervision would have to be assessed under the law of simple agency contract. The Federal Court's view has therefore been criticised: according to Gauch (*Architekturvertrag*, nn.38ff.), the blanket contract constitutes a simple agency contract.

If we adopt the Federal Court's view, the error which the Architect made in the present case would probably rate as a design error, and so the law governing work contracts would apply. Reference can here be made, *mutatis mutandis*, to the comments on the nexus between developer and contractor under work-contract law. If, however, we adopt the view taken in the doctrine, which places the blanket contract under the law of the simple agency contract (Gauch, *Architekturvertrag*, nn.38ff.; Schumacher, *Haftung des Architekten*, n.397, footnote 58), the Architect's liability is governed by the following rules.

1 Law sources (law governing simple agency contracts)

1.1 By SCO S.394, the agent under a simple agency contract undertakes to carry out as agreed the business committed to him.

1.2 The agent will be liable to the principal for faithful and diligent execution of the business committed to him (SCO S.398, para.2). As to the degree of diligence to be observed by the agent, paragraph 1 of that section makes reference to the relevant provisions of labour law.

1.3 The relevant liability is particularised by SCO S.321.e: The employee is liable for any loss which he wilfully or negligently causes to the employer. The degree of diligence required of the employee (and, through the reference made, of the agent) depends on the given employment (or agency) contract, in having regard to the occupational risk, the education level or technical knowledge required for the work, and to the employee's (or agent's) abilities and characteristics of which the employer (or principal) was aware or ought to have been aware. Doctrine, however, postulates that the diligence and loyalty duties of a free-lance architect should be subject to stricter requirements than those of an employee (Schumacher, n.436).

1.4 A first condition for the Architect's liability is that he breached the particular architect's contract. Thus, he may have failed in general diligence duties. Such duties include the duty to warn (Schumacher, nn.449ff.). The warning must be given expressly and must unmistakably inform the principal that the instruction is misconceived, and that observance thereof might jeopardise the principal's interests. In particular, the Architect must inform the principal of the risks inherent in the instruction, e.g. the risk of using new building materials which, though desired by the principal, have not performed well in practice (Schumacher, n.475).

1.5 The architect's breach must be shown to have caused loss to the principal.

1.6 There must also be an adequate causal connection between the architect's conduct and the loss.

1.7 The architect must have acted culpably; negligence is sufficient. It is for the architect to prove the contrary.

1.8 If the foregoing conditions are present, the architect must indemnify the principal.

1.9 The principal's claims lapse after ten years.

2 Substantive issues

2.1 As mentioned, the documents available do not specify the terms of contract between developer and Architect. We will therefore assume that the degree of diligence incumbent on the Architect is to be decided under statutory rules.

2.2 As regards breach of contract by the Architect, the crucial point in the present case is whether he gave proper warning. In our opinion, it is not sufficient for the Architect to point out that he has no experience of the particular material; in that case, he would have been bound to obtain the necessary information. A relevant paper was indeed available at the time, indicating the unsoundness of the tiling. To the developer, the Architect's statement that he had no experience of the material did not necessarily mean that the tiling was unsound. The developer was entitled to expect that if there were any reasons against using that tiling, the Architect would have told him so, and that the Architect would do what was necessary to find out for certain. Yet, that is precisely what the Architect did not do. The Architect therefore breached the Architect's contract by failing to give proper warning.

2.3 A further condition for the Architect's liability to indemnity is that loss actually occurred (cf. Schumacher, nn.527ff.). The impaired outward appearance of the building has considerably diminished its value. Moreover, the Bank has incurred loss of rent. It has therefore incurred both property damage and financial loss.

2.4 The condition of adequate causal connection implies that the breach of contract and the loss are so connected that by the general run of things and by general experience of life, the breach is apt to bring about the particular loss (Schumacher, n.574). That condition is likely to apply in the present case.

2.5 By the general rules of the SCO the Architect will not be liable unless he is proved culpable; but it is for him to prove that he is not (SCO S.97, para.1 and S.402, para.2).

Negligence is sufficient. An architect must live up to all the capabilities expected of him when he acts in an architect's capacity (Schumacher, n.578). Where he lacks experience of his own, he must obtain specialist advice or enlist a specialist. So he would have been bound in the present case to make a close inquiry about the material desired by the developer, and such a close inquiry would most likely have led him to the paper published by the building research institute. Culpability on the Architect's part is therefore to be affirmed in the present case.

2.6 Once the Architect is found liable, he must indemnify the Bank for the loss incurred. The form of indemnity is decided by the court under SCO S.43 para.1. Damages are the rule (Schumacher, n.593). The claim to damages is limited by the actual loss. Yet, the loss proved cannot readily be recovered in full. In a dispute, the sum will be determined by the court (under SCO S.43, para.1). The court will consider the degree of culpability, the circumstances of the case, and any degree of culpability on the claimant's part (SCO S.44, para.1). True, the claimant, i.e. the Bank, did not give any instruction and is therefore not to blame; but as it can only claim against the Architect on the strength of the assignment made by the developer, the Bank must accept the developer's culpability, if any, as its own. In that light, the court is likely to reduce the award of damages.

2.7 Neither agency-contract law nor the general law of liability for breach of contract recognises any inspection or notice-of-defect obligations of the claimant involving loss of rights (Schumacher, n.631). The Bank can therefore sue the Architect any time before expiry of the ordinary limitation period. By SCO S.127, the Bank's claims lapse ten years after the breach of contract (Schumacher, n.642).

The Bank has a loss-reducing duty. Thus, it may be bound to notify the Architect of the defect immediately on discovery if such early notice is apt to prevent loss or prevent further loss. If the Bank fails to notify the architect immediately, it will have to bear such part of the loss as could have been prevented by early intervention. The documents available do not reveal whether such immediate notice might have helped reduce the loss.

2.8 If the SIA Standard 102 was adopted by the parties, its terms as to notice period and limitation period will apply. By SIA 102, head 1.8.2, third clause, notice of defects in an immovable building must be given immediately. Yet, the consequences of omitting to do so are not regulated. We will therefore assume – with due reservation – that the said clause does not alter the statutory rule (Schumacher, n.637).

The relationship between architect's liability and contractor's liability in respect of the investment bank's claims

If the conditions for liability are present both for the contractor and for the architect, both will be liable for the defects in the building (Gauch, *Werkvertrag*, n.274). The investment bank (as successor in title to the developer) can therefore sue both as liable. The contractor's liability for defects and the architect's liability in agency contract subsist concurrently in equal rank. The provision applicable is SCO S.147 para.1, whereby either debtor is released from satisfying the creditor to the extent that the other has given satisfaction (Gauch, *Werkvertrag*, n.2746). The double liability must not result in enrichment of the Bank. Whether either debtor may recover from the other depends on the allocation of liability between the two; the court may decide the point within its discretion (Gauch, *Werkvertrag*, n. 2748). The documents available do not give any particulars on which to venture an opinion on the allocation of liability between architect and contractor.

THE INVOLVEMENT (IF ANY) OF INSURERS IN THE PURCHASER'S CLAIM FOR RECOVERY, INCLUDING THE METHOD OF SOLVING THE MATTER

1. If the developer as vendor has a liability policy, the insurance will not cover the investment bank's claims in rescission or diminution: Such claims relate entirely to improper performance of contract. If the developer is sued for consequential loss, the benefits, if any, will depend on the given insurance contract. Such contracts normally rule out benefits for purely financial loss.

2. If the contractor has a business liability policy, the following will apply:

The 'Warranty Clause' contained in the 'General Terms of Contract' excludes claims for performance of contract or alternative claims for indemnity for non-performance or improper performance. That applies notably to defects in, and damage to, products produced by the insured (Hepperle, pp.212f.). Loss consequent on such defects will be covered by the insurance only if damage to persons or property is involved. The documents available mention nothing of that. We will therefore assume that the contractor's liability insurance will not cover the investment bank's loss because proper performance of contract was the contractor's own concern.

3. If the contractor has a construction policy, the following will apply:

The construction insurance covers construction work (and further objects if applicable) to be provided by the contractor, against damage or destruction caused by construction accidents. So the insurance will not cover any cost which the insured incurs for the removal of defects due to faulty execution or the use of unsuitable materials (Hepperle, p.202). Yet, precisely such loss is involved in the present case. The contractor will therefore not be able to recover the cost of remedial work from the construction insurance.

4. If the architect has a professional liability policy – which is the rule – the following will apply:

The professional liability insurance as a rule also covers what is called 'buildings damage', i.e. damage to, and defects in, buildings for which the insured is liable as a professional (Schwander, n.1887; Hepperle, p.211).

The insurance thus also covers defects in buildings due to an error made by the architect. Loss consequent on defects is placed on the same footing as the defects themselves.

In the present case, it means that the architect's professional liability insurance would probably cover the loss which the investment bank incurred and for which the architect is answerable, though it would doubtless not cover the full loss.

THE PROCEDURES BY WHICH THE ABOVE ARE RESOLVED, INCLUDING EXTRA-LEGAL, LEGAL AND INSURANCE PROCEDURES

As regards the procedure by which the Bank would have to claim, it is to be noted that the law of civil procedure in Switzerland is a matter for the Cantons. The rules governing jurisdiction (as to subject matter) and proceedings therefore differ greatly from one Canton to another (Hürlimann, *Dispute Resolution*, p.730)

Where insurance is involved, and if the claims of the parties are liquidated, disputes will be settled by negotiation between the insurance companies.

Type B–I

The *Hotel* case

Viktor Aepli

1 The Plaintiff

The plaintiff was a private corporation running at least one large hotel. As such, it would occasionally commission construction work.

2 The Works
The works consisted in sealing and lining the balconies and terraces of the Plaintiff's hotel.

3 Procurement Method
In 1989, the Plaintiff entered a turnkey contract with a building contractor (the 'contractor'). The works would be considered completed when the terraces and balconies were ready to use.

4 Insurance
The contractor had insurance cover, with a fixed upper limit.

5 Damage or Loss
Rainwater seeped under the new sealant and linings, resulting in blistering which was detrimental to the appearance of the building and rendered the normal use of the balconies and terraces impossible.

The damage was discovered by an employee of the Plaintiff.

10 Process after Discovery
The Plaintiff immediately demanded substantial repairs. When the contractor refused to carry them out, the Plaintiff demanded monetary compensation, including consequential damages, to the degree necessary to allow the normal use of the balconies and terraces.

The Plaintiff then filed a complaint, naming the contractor.

The contractor informed its insurer of the claim. In line with the terms of its policy with the contractor, the insurer paid the Plaintiff the fixed upper limit, which was insufficient to cover the damages claimed. The Plaintiff initiated a claim against the contractor in order to recover the remaining amount.

When the lower court had handed down its decision, the contractor appealed.

11 Technical Reports
Both parties commissioned expert reports. All reports concluded that the contractor was responsible for the damage incurred; these reports were not accepted by the contractor.

However, one of these reports, prepared by a jointly appointed and agreed technical expert, was commissioned on the basis that it would be a *Schiedsgutachten* (a binding report by a technical expert), which under Swiss law is binding upon the parties as well as upon the court.

20 Resolution

The matter was resolved in a lower court, and, upon the contractor's appeal, confirmed in the Appeals Court. At the time of writing, it was still possible that the case would proceed to the Federal Court, the highest court in Switzerland.

21 Costs

According to the lower courts, all costs (fees as well as costs incurred by the parties) were to be paid by the contractor; these included the cost of the District Court's official confirmation of the evidence in anticipation of the trial, which was incurred long before the trial.

22 Information

Both parties had full access to all information.

23 Time

The sealing and lining work was, as mentioned, commissioned in 1989. The lower court heard the case in 1993 and issued its decision in 1996.

As mentioned, the contractor appealed, and the Appeals Court's decision, confirming that of the lower court, was handed down in 1996.

24 Role of Experts

As mentioned, the Plaintiff submitted various technical reports as evidence in the trials in both courts, including the report commissioned by both parties on the basis that it would be binding.

Neither court ordered additional technical reports. However, at the time the damage occurred (and long before the case was heard in court), the Plaintiff obtained an official report from the local court, issued in anticipation of future litigation, which set out the extent of the damage.

The Plaintiff presented this report as evidence during the first trial. On the grounds that the contractor had not been a party to this anticipatory confirmation of the evidence, the court found it to be a mere assertion by one of the parties, rather than an objective and official statement of the facts.

During the first trial, the author of one of the technical reports (but not of the *Schiedsgutachten*) testified as an expert witness.

25 Outcome

Both courts found in favour of the Plaintiff. The decision of the Appeals Court turned on the fact that before the trial began, the parties had agreed to commission a *Schiedsgutachten*, and that this had concluded that the contractor had not performed to the standards of the industry (the *Schiedsgutachten* found that the materials used by the contractor were inappropriate, given the condition of the balconies and terraces).

Under Swiss law, the judge is not permitted to overrule the conclusions of a *Schiedsgutachten* commissioned by both parties unless the report is found to be entirely untenable, i.e. obviously unjust, arbitrary, careless, full of errors, or based on a false description of the facts. A judge may not overrule a *Schiedsgutachten* on the basis of mere doubt as to its correctness: the judge must be wholly convinced that the expert committed an obvious and serious mistake. The Appeals Court found that the *Schiedsgutachten* in this case was not vulnerable to such criticism, and therefore was binding upon the parties. This being the case, the other reports submitted as evidence by the Plaintiff and the testimony of witnesses during the first trial were *a priori* irrelevant.

The contractor's objection that the author of the Schiedsgutachten was not competent was overruled by the Appeals Court.

The court found that the contractor failed to object to a legally relevant error in the agreement by which the *Schiedsgutachten* was commissioned within the time allowed by law.

30 Feedback
This case is not available to the public.

Type B–II

The *ABC* case

Roland Hürlimann

1 The Plaintiff
The plaintiff was a state institution.

2 The Works
The project was a public building with difficult soil conditions.

3 Procurement method

This was a traditional design-and-build project, with a joint venture between contractors.

4 Insurance

All contractors in the joint venture carried their own insurance covering their work, plus a separate cover for the joint venture's activities.

5 Damage or loss

Several neighbouring buildings were affected by cracks.

The damage was identified by experts.

The cause of the damage is still not identified beyond dispute.

10 Process after Discovery

The Client filed a complaint directly with the Federal Court of Switzerland, the country's Supreme Court, naming the construction companies' joint venture, and claiming compensation for the cost of repair works to neighbourhood buildings, plus consequential damage.

The construction companies' joint venture informed its insurer of the claim.

11 Technical Reports

Both the plaintiff and the defendant commissioned several expert reports, none of which was accepted by the other side.

In its complaint, the plaintiff asked the court to nominate a judicial expert to identify the cause of damages.

20 Resolution

The matter was still not resolved at the time of writing.

21 Costs

Costs have not yet been apportioned.

22 Information

Both parties have full access to all information.

23 Time

The complaint was lodged in 1994 with the Federal Supreme Court in Lausanne and was not settled at the time of writing (1997).

24 Role of Experts

As mentioned, both the plaintiff and the defendant submitted various technical reports as evidence in trial. The Federal Supreme Court is nominating another expert.

25 Outcome

The case, still pending, will be made available to the public once settled.

Literature quoted:

Gauch, Peter: 'Der Werkvertrag', 4th edition, Zürich, 1996

Gauch, Peter: 'Kommentar zur SIA-Norm 118, Arts 150–90', Zürich, 1991.

Gauch, Peter *et al*: 'Kommentar zur SIA-Norm 118, Arts 38–156', Zürich, 1992.

Hepperle, Erwin: Bauversicherungen in Martin Lendi *et al*: Das private Baurecht des Schweiz, Zürich 1994.

Hürlimann, Roland: Der Architekt als Experte in Peter Gauch/Pierre Tercier eds: 'Das Architektenrecht', 3rd edition, Freiburg 1995.

Hürlimann, Roland: 'Dispute Resolution and Conflict Management in Construction', Swiss Report, pp.729–93, London/New York 1998.

Schumacher, Rainer: Die Haftung des Architekten aus Vertrag, in Peter Gauch/Pierre Tercier eds: 'Das Architektenrecht', 3rd edition, Freiburg 1995.

Schwander, Werner: Die Haftpflichtversicherung des Architekten in Peter Gauch/ Pierre Tercier eds: 'Das Architektenrecht', 3rd edition, Freiburg 1995.

United States

Wilson Barnes and Ron Craig

Type A

INTRODUCTION

We shall approach the following discussion in a general way with specific features of jurisdictional reference applied as appropriate. Every State in the United States is a separate jurisdiction, and the references will be chiefly to the State of Florida, with some other States referred to as appropriate.

Under the conditions as described, the investment bank (the Bank), has decided that hairline cracking of cladding tiles on that building is deterring potential lessees.

It is normal procedure in the US for a building appraisal to be conducted by or on behalf of lending bodies. Appraisals normally include an inspection component, and this may or may not be the missing inspection referred to in the scenario. Although there is the possibility of a negligent appraisal, if we set aside issues of the Bank's failure to inspect the building before purchase and the absence of any construction professionals on the developer's staff, we can move to the substantive matters of liability.

At the outset, we note that the Bank is a remote party to the original building development, including its financing and construction. As such, unless expressly agreed to otherwise, the Bank is not entitled to the normal prerogatives associated with privity of contract in terms of designer or contractor liability. Secondly, the tiles appear to be a nominated product, identified by the developer and, specifically, nominated over the protests of its architect. Third, although the tiles had been identified by the building research institute as not suitable for exterior use, it is not clear that the

Case Studies in Post-Construction Liability and Insurance, edited by Anthony Lavers. Published in 1999 by E & FN Spon, 11 New Fetter Lane, London EC4P 4EE, UK. ISBN: 0 419 24570 7

architect, and thus the developer, were aware of this, or if any dialogue took place between them and the manufacturer prior to specification or construction. Also, it is not clear at what point in time the tile manufacturer knew of the tiles' tendency to crack, and, further, whether the referenced paper on the issue was available before or not until after the specification was written. Fourth, we are dealing with an economic loss issue as opposed to one of personal injury or property damage either to the building itself or to other related real property. Fifth, the matter of warranty on the tiles is an open issue since we have no information on that point.

'Collateral warranties', in the sense described by Jenny Baster *et al* in their discussion of England, Wales and Northern Ireland law on this point[1] do not occur frequently or in such a classified manner in the US. 'Unless the specific terms of the contract between the client and the producers or the terms of transferable warranties issued by the producers to the clients provide otherwise, successive owners are generally not entitled to recover damages from producers.'[2]

Florida law embodies a 'Collateral Source Rule', which in essence recognises that claims of contributions against damages in negligence suits may be made with the defendants naming collateral sources of the contribution. The permissibility of such claims is regulated on the basis of the proposed contributor's right to subrogation; the lack of such a right allows a *bona fide* reduction of the damage recovery,[3] while the presence of a right has been found to preclude the inclusion of the claimed sum in the account of payment against the prescribed damages.[4]

We shall address the above five features – privity, nominated product, awareness of manufacturing problem, economic loss doctrine and warranty. The reader should bear in mind that every State in the US is a separate jurisdiction that is overlaid by the federal laws. We shall respond to the provided questions primarily in accord with the State of Florida laws, and draw from other jurisdictions when appropriate citations to amplify our arguments.

THE PURCHASER'S OPPORTUNITY FOR RECOVERY, THROUGH LITIGATION OR OTHERWISE, FROM THE VENDOR

Opportunity for recovery could flow through several paths: privity, specific conditions of sale, statutory warranty (merchantability or fitness) and original warranty by the manufacturer. Post-construction sales of property are normally on an 'as is' basis, with exceptions or obligations of the seller carefully spelled out and acknowledged by both parties to the contract.

Standard sales of real estate are governed by purchase-and-sale agreements, which in effect are contracts for ownership exchange. Just as terms of the contract are important in agreements between clients, designers and contractors, they are important in transfers of ownership. If the sales contract terms provided for the purchaser to have recourse for the tiles, if, as they did, they should develop visual flaws or some similar problem, then the purchaser may have a right to recover from the vendor. Without such a provision in the sales contract or manifest and well-founded belief in the good faith of the seller, the purchaser is as vulnerable as P.T. Barnum's 'There's a fool born every minute.'

As noted in *Florida Construction Law Manual,*

> If the purchaser has a reasonable opportunity to discover a defect prior to sale and fails to exercise that opportunity, then failure to discover the defect can be a defense to a breach of warranty action.

Leiby cites *Putnam v Roudebush*[5] in support. Absent other qualifying conditions, the Bank's failure to conduct any kind of property inspection prior to purchase leaves it in a very weak position for any claim of building imperfection.

Normally, any warranty on the tiles – other than one specifically described as transferable to future owners and for an extent of time covering the purchaser's complaint – would not benefit future owners. It is arguable that the statutory warranty of fitness of goods (Fla. Stat.§ 672.315 (1965)) and the statutory warranty of merchantability (Fla. Stat.§ 672.314 (1965)) might be pertinent. These both apply to sales of goods, and while merchantability means that goods must be of average quality and perform as advertised, fitness assumes that the seller is aware of the intended use and knows that the buyer relies on the seller to provide suitable goods. Both concepts are embodied in the Uniform Commercial Code, a federally generated instrument adopted voluntarily by all 50 states (Leiby, § 12:09).

In the US, there are express and implied warranties:

> An express warranty is the specific written or oral representation which distributes the risk of specified defects or failures between the parties to an agreement. An implied warranty is the guarantee of one party against failure or defect in a product or transaction which is imposed on that party by operation of law or which is inferred from the actions of the parties.[6]

In discussing construction warranties and the Uniform Commercial Code, Stein says clearly that express warranties do – and implied warranties of the UCC do not – extend to future performance. He cites numerous cases in support of this and notes succinctly that '[t]he drafters of the UCC intended to reserve the benefits of an extended warranty to those who explicitly bargained for them.'[7]

Specification by the buyer of a particular brand is noted generally as negating reliance on the manufacturer's knowledge and expertise, but in UCC provisions is expressly not a *per se* exception to the warranty of fitness.[8]

The flaw in the tiles has caused the new owner economic loss; there is no personal injury or property damage involved here. Leiby notes that

> The implied warranty of merchantability is not a proper legal theory for recovery of economic loss (as distinguished from personal injury or property damage) without privity of contract.'[9]

He cites numerous cases in support of this and notes further that while in the past there were instances of findings for breach of implied warranty in the absence of privity, they were impliedly overruled by the Florida Supreme Court in 1993.[10] Similarly, he notes (§ 12:09) that an 'action for breach of an implied warranty pursuant to performance of a contract does not lie where there is no privity of contract,' and refers again to the above cited Florida Supreme Court ruling. Roughly half of the States in the US require privity for economic loss actions based on negligence, and roughly half do not; there is no country-wide consensus on this issue.

Written warranties are usually valid for at least a year, sometimes up to five and occasionally to 10 (e.g. some roofing materials). Warranties on goods are normally predicated on their installation according to manufacturer's instructions. Such warranties can survive a transfer of ownership, depending on the specific wording of the warranty as issued.

In the case under study, under the most likely set of relationships – i.e., a lack of privity between the Bank and the producers of the building –, such absence of privity will preclude third-party recovery for economic loss, apart from the most favourable written warranty still in effect at the time of notice and providing for restitution, replacement or repair.

Some interesting language involving economic loss and a building product, which did not perform properly, arose in a case where an architect who purchased Cor-Ten steel panels was barred by the economic loss doctrine from recovering tort damages from the manufacturer. Cor-Ten is supposed to rust superficially and then form a protective coat that resists further corrosion, without the need for paint; in this case, the panels rusted through and required replacement. The buyer was restricted to an action for breach of warranty.[11] The court stated that

> It would be better to call it 'commercial loss' [economic loss], not only because personal injuries and specialty property losses are economic losses,—they destroy values which can be and are monetized—but also, and more important, because tort law is a superfluous and inapt tool for resolving purely commercial disputes.[12]

In general, the courts do not regard favourably the concept of damages to redress economic loss. This is a stance which is found in the law of tort in other common law systems also. Many past favourable judgments have been superseded by subsequent denials, and even insurance language has come to reflect the trend. This is exemplified in an insurance type called *Completed Operations Insurance*, whose standard-language forms have experienced successive revisions to exclude provisions that favoured recovery in instances of economic loss. This is discussed below.

THE PURCHASER'S OPPORTUNITY FOR RECOVERY, THROUGH LITIGATION OR OTHERWISE FROM THE VENDOR'S CONTRACTOR

The major arguments set out above are equally applicable here: privity, specific conditions of sale, statutory warranty (merchantability or fitness) and original warranty by the manufacturer. Economic loss continues to be the governing characteristic that identifies opportunity options for recovery in this case. No other characteristic has been mentioned (we assume that there is no damage to person or property other than what has been described). Should the Bank argue negligent conduct on the part of the contractor, it would have no legal basis for an action in Florida: 'Damages of economic loss are not recoverable under a negligence theory absent privity of contract.'[13]

The contractor in Florida has a basic duty of care to produce a building in a workmanlike manner. This does not mean perfection or that extremely high quality is required, only work that is consistent with the accepted norm. This concept is regarded as an implied warranty, equivalent to an implied warranty of merchantability.[14] Typically, contractors act as brokers and use sub-contractors for portions of the works. The following case illustrates a privity relationship involving a nominated product and implied warranty.

In a 1970 Florida decision,[15] a sub-contractor agreed to furnish materials of good quality, unless otherwise specified. The sub-contractor installed *Miami stone*, as directed by the specifications. This particular type of brick was defective. By requiring the sub-contractor to furnish brick of only a certain distinctive type produced by a single manufacturer, the contractor relieved the sub-contractor from its obligation to furnish only brick of good quality (Acret, pp.311–2).

In a similar vein, contractors are liable for an implied warranty that buildings are constructed according to plans on file with local building authorities (Leiby, § 12:06). But it is also mentioned by Stein in a discussion of suitability under the implied warranty features of the *Spearin* doctrine that

> The general rule is that when a contractor follows plans and specifications supplied by the client, which later prove to be defective or insufficient, he is not responsible to the client for loss or damage resulting therefrom as a consequence of the defectiveness or insufficiency of such plans and specifications.[16]

While a contractor has a primary concern to complete his contracts in time and consistent with regulatory requirements and safety, he is subject to suit arising from the construction of his projects. The vulnerability period varies from State to State in the US; in Florida, four years are allowed for the filing of an action related to 'design, planning, or construction of an improvement to real property.' In the main, the time runs from the latest date of: possession by the client, abandonment of construction, or termination of contract between the contractor and its client. But in the case of a latent defect, the time runs from when the defect is discovered or should have been discovered with due diligence. This is known as the *Statute of Limitations*. Under any circumstance, action must commence within 15 years from the latest of the primary date options listed above, and this feature is referred to as the *Statute of Repose*. These two concepts are both defined in the same section of the Florida Statutes (§ 95.11(3)(c)(1995)). The basic four-year period cited above is further specified as two years in professional malpractice actions (other than medical) for parties in a relationship of privity of contract with the professional (§ 95.11(4)(c)(1995)).

Assuming a proper tile specification by the architect, and improper fulfilling of that by the tile supplier, a subsequent agreement between the new owner and the contractor to replace the tile could benefit from a 1969 Florida decision.[17] In this case,

> a supplier agreed to supply roofing materials to a sub-contractor on an Air Force project, even though the supplier knew that the material failed to meet specifications. The prime contractor was entitled to bring suit against the supplier as a third-party beneficiary of the contract between the sub-contractor and the supplier. The prime contractor was the one that purchased replacement material for the sub-contractor. Therefore, the prime contractor was the only one that sustained the loss, and the only one that should be able to recover from the supplier to avoid unjust enrichment.[18]

This appeared in a chapter on rights against remote parties, which treats third-party beneficiaries at length. Acret notes that

> although the reach of the third-party beneficiary doctrine is expanding, it is still the general rule that contracts only create enforceable obligations between the parties to the contract. A third party who derives incidental benefit from the performance of a contract has no direct right to enforce the contract.'[19]

THE PURCHASER'S OPPORTUNITY FOR RECOVERY, THROUGH LITIGATION OR OTHERWISE, FROM THE VENDOR'S ARCHITECT

Architects in general have a duty of care as licensed professionals. In Florida, 'the standard of care applicable to an architect or engineer is compliance with standards of good practice recognised at the same time and in the same locality.'[20]

The nature of information on the suitability of the nominated tiles available to the architect and the client at the time of final specification is of considerable importance here. We assume that despite the architect's discomfort and desire not to use the tiles, he was not aware of the formal statement of unsuitability from the building research institute. If he was, and failed to document the client's instruction to use the product, then he could be liable for negligence. On the premise that he was not aware of any formal warnings and had merely advised against the tiles on some lesser grounds, the client was in a position of having bought what he asked for.

It is also possible that the manufacturer did not know of the latent defects in the tiles and was unable to warn, or he did know and was guilty of misrepresentation as to the fitness for intended purpose of the product.

The Kansas Supreme Court applied an earlier ruling to an appeal of a case involving roofing product unsuitability for specific application:

> No manufacturer, ruled the Court, has an absolute right to disseminate false information about its products, knowing that others will rely on that information to their detriment. Similarly, a manufacturer who knows that its product is being improperly installed or used cannot continue to disseminate with impunity information encouraging the improper installation or use of such a product. Moreover, the manufacturer is required to keep abreast of the current state of knowledge as gained through research, adverse reaction reports, and other available methods. *Wooderson v Ortho Pharmaceutical Corp.*, 681 P.2d 1038 (Kan 1984).[21]

Continuing with the case, the Court noted that the manufacturer 'knowingly and wilfully led another party' into constructing an unsuitable and defective roof, which could have been avoided by withdrawal of the faulty specifications. The Trial Court's award of punitive damages against the roofing manufacturer was reinstated. The Kansas Supreme Court ruled similarly in *Stephen v GAF Corporation.*[22]

Leiby refers to a Florida case in defining fraud and says (§ 16:02): 'fraud is the intentional misrepresentation of a material fact, which misrepresentation is relied on and causes damage.' This is straightforward and readily understood. In his following definition of negligent misrepresentation, he includes those elements that he ascribed to fraud, except that negligent misrepresentation seems to be distinguished by absence of the word 'intentional' (§ 16:03). Acret offers some explanation on this:

> In many circumstances, courts recognize fraud that does not coincide with the conventional definition. Fraud may occur where the misrepresentation is merely negligent, or even *innocent*, and also may be found where the damaging statement was technically and literally true.[23]

The statutes of limitation and repose described above are applicable to architects and engineers as well. The primary options for time to start running include the date of termination of contract between the professional and his client (Fla. Stat. § 95.11(3)(c)). The basic covenant, mentioned in earlier sections, of no viable action based on negligence in the absence of privity between the parties is also applicable here in the issue of opportunity for recovery from the architect.

THE INVOLVEMENT (IF ANY) OF INSURERS IN THE PURCHASER'S CLAIM FOR RECOVERY, INCLUDING THE METHOD OF RESOLVING THE MATTER

It is standard practice for contractors to carry a variety of insurance covers. Not only is this frequently a local requirement, but the increasing level of litigation in the industry has caused many to seek insurance protection they might earlier have ignored. Despite the need for contractors to forge alliances with insurers, there is an underlying suspicion of the cover that insurance blankets actually provide. This is well expressed in Acret's introductory section to his chapter on insurance.

> It would be an appalling task meticulously to lucubrate, parse out, and fully explain the true meaning of an insurance policy. The language employed by these documents is arcane, often only to be understood after a painstaking study of the history of the development of the business of insurance. Insurance policies exemplify all the characteristics of contracts of adhesion: they are standardized contracts, drafted by a party of superior bargaining power, offered to a party of inferior bargaining power, who is given no opportunity to negotiate the language. Indeed, as courts recognize, the idea that consumers of insurance read their policies (at least before a loss occurs) is a flagrant fiction. It is not surprising, therefore, that insurance policies are, and rightly should be, interpreted liberally in favor of the insured (Acret, pp.343–4).

Such liberal interpretation was exemplified in a 1965 case in Florida holding the insurer of an electrical sub-contractor liable for damage to work he had completed. Despite a policy exclusion clause for injured property in the *care, custody, or control* of the insured, the completed work found damaged had been in a locked room for which an employee of the sub-contractor did have a key. The employee was found to lack the access requisite for the inherently ambiguous terminology of this exclusion to become operative (ibid. p.344).

The general perception that insurance policies are difficult to read and understand (sometimes made so deliberately) is given recognition in Florida where policies are required to be easily readable (Fla. Stat. § 627.4145(1992)). That State has an insurance code found in Chapters 624–632 of the State statutes (Leiby §20:01). Florida has extensive regulations governing insurance agents as well as the policies that they represent.

Both clients and contractors carry insurance during the course of a project's lifetime, i.e. from inception to hand-over or occupancy. Thus both must consider multiple risks and requirements to insure against them.

Increasingly, contractors carry a comprehensive policy that bundles many of the covers together. Invariably, this cover terminates with hand-over or occupancy or termination of the contract. Protection beyond these times can be obtained under a concept called *Completed Operations Cover*:

> Besides the possibility of liability arising from bodily injury or property damage during current, on-going operations, contractors and others[24] in the construction arena face a perhaps larger risk of bodily injury or property damage arising from their work after it is completed, often long after it is completed. This is known as the 'completed operations hazard.'[25]

Completed Operations Cover is available as a feature of most project liability policies. In many cases now, it appears as an exclusion that may be removed by additional premium. Leiby refers to this cover as '...analogous to products liability cover for manufacturers of goods,' and cites support for this position.[26] However, his discussion of the concept and citation of various relevant cases focuses on instances of bodily injury or property damage. Philosophically, this parallels the attitudes of no damages for economic loss in negligence in the absence of privity of contract.

Acret (pp.358–9) refers to an Arizona case from the Ninth Circuit Court in 1973.[27] Following a roof collapse due to heavy snowfall, the insurer refused to honour a breach-of-warranty-and-negligence claim by the owner. Summary judgment found a valid policy exclusion for damage to work performed by or on behalf of the insured contractor arising out of the work. Cover was also excluded for property damage to the insured's products arising out of those products. Despite a Completed Operations Cover in force, it was limited by the exclusions.

Back in Florida, we find (Acret p.359) a decision related to a collapse of a wall killing a young child in 1973 which illustrates the ambiguity of exclusion language:[28]

> The policy was held at best ambiguous, since it was stated to cover *hazards* including the ownership, maintenance, and use of premises *and all operations*. The exclusion of *completed operations* was rendered ambiguous by the inclusion of cover for hazards arising out of all operations.

The ambiguity was resolved in favour of the insured.

We have referred several times to work-products exclusion. Since 1973, such language appears in many comprehensive general liability insurance policies. The wording is generally that mentioned above relative to *Southwest Forest Indus v Pole Bldgs, Inc*. There is a so-called broad form of exclusion that deals with the same issues, but in slightly different language, used by some insurers. The major difference is less restrictive exclusion in the broad form, frequently allowing cover for work of sub-contractors. Losses described in terms other than those identifiable as falling within the broad form or regular form phraseology may overcome the exclusions, e.g.:

> If defective plaster is applied to a building, there are two ways of defining the damage: (1) the plaster is work performed by or on behalf of the insured, the damage is to the work, and therefore the loss is excluded, or (2) the market value of the entire building is damaged, the entire building is not work performed by or on behalf of the insured, and therefore the damage is covered.

Those were the essential facts in a 1954 Minnesota Supreme Court case.[29] It was held that the value of the building was diminished by the presence of the defective plaster; therefore, the loss was covered by the policy. Such was also the holding in a 1984 Eighth Circuit decision under Missouri Law.[30]

In a 1984 California decision,[31] an insured supplier contracted to sell 90,000 board feet of specially milled siding to a developer. The supplier in turn contracted with a milling company to process and deliver the lumber. After 25 percent of the siding had been installed, it became apparent that it had been mismilled: the pieces were obviously unsightly. The work-products exclusion applied to the siding itself, but did not prevent cover for damages to the eight houses, which were diminished in value as a result of the application of the defective siding.

In a 1966 Third Circuit case interpreting Pennsylvania law,[32] a manu-facturer produced 'Steelbestos' (a metal siding bonded to asbestos sheets). After the siding had been installed in several buildings, it began to delaminate and became discoloured. The insured made extensive repairs, incurring expenses of USD 417,000, excluding the cost of manufacturing new siding. Once the siding was installed, the building suffered a diminution in value covered by the insurance policy.

Acret (who quotes the cases above on p.361), also notes (on p.367) that diminution in property value resulting from incorporation of defective product(s) is not covered in the current (as of 1986) comprehensive general liability policy language. The fine print in policies is to be read and understood. During the past several decades, considerable debate has occurred over the meaning of the so-called 'product exclusion' clauses in the standard general liability policies used to insure contractors' activities.

Standard policies are formatted by the Insurance Services Office, Inc. and used by various insurance companies subject to rules of the State jurisdictions. The general liability policy applicable to construction has been known as the *Comprehensive General Liability* since 1973. It is still available in the older format but was also upgraded in 1986, as *Commercial General Liability* policies. These 1986 policies are in two forms distinguished by the trigger, namely either 'claims-made' or 'occurrence.' The previously mentioned 'work-product' exclusions of these policies account for some of the most important and controversial language in construction insurance. No cover for: damage to property under 'care, custody or control' of the insured; damage to property the insured is working on; damage to the product or property 'arising out of it or any part of it' with respect to either work in progress or completed operations. Collectively,

> These typical exclusions have worked as a composite to exclude cover for damage to the contractor's work arising out of defective construction. The broad form endorsement to the old comprehensive general liability policy, and the new commercial general liability policies as written, do provide the contractor with cover for damages to completed operations arising out of a sub-contractor's work.'[33]

The editing of exclusion clauses, among other language in successive publications, can be said to have influenced the views of the courts and thus the state of the common law concerning the liability of contractors for the products of their efforts. Standard policy language upgrading is also responsible for seemingly contradictory findings in matters of 'completed operations' and 'additional insureds' which are frequent issues in construction cases. See S.C. Turner, Chapters 26, 32 and 40 for a full and sufficient explanation of the subtleties involved in these matters.

THE PROCEDURES BY WHICH THE ABOVE ARE RESOLVED, INCLUDING EXTRA-LEGAL, LEGAL AND INSURANCE PROCEDURES

There is a variety of ways for solving the above, ranging from the adversarial confrontation to informal fact finding and negotiation conferences. In many situations, insurance representatives make themselves readily available to their clients with personal advice and factual guidance from massive insurance industry databases. These clients may be owners, clients, designers, contractors and other parties.

Despite the inclination for the insureds to think that insurers are on their side, the literature contains many references to courts holding that insurers have 'a duty' to defend the insured, thus forcing the insurers to participate and sometimes pay damages.

Over the years, construction contract forms and insurance policy forms have been periodically upgraded. New forms invariably contain new wording that reflects the negative experiences of the major parties, i.e. the owners, the clients, the designers, the contractors and the insurers. Each group attempts to strengthen the language barriers protecting its own interests, yet a certain negotiation goes on among all. Court decisions frequently play a role in establishing a normative behaviour, that is, normative until it is superseded by more current thinking. For instance, in New Hampshire, where 'naked exculpatory provisions' are not favoured, the court has approved exculpatory terms included in a construction contract as 'part of a larger comprehensive approach to indemnifying the parties involved ... allocating the risks involved, and spreading the costs of different types of insurance.[34]

Many of the cases cited here either have resulted in revision of forms and procedure or will do so in the future. The same can be said with regard to the development of statute and common law, which generally precedes actions and counteractions, by the other participants.

ANY MECHANISMS EXISTING FOR UTILISING THE FEEDBACK OR EXPERIENCE FROM THE CASE TO A BODY OR BODIES WHICH MAY MAKE USE OF IT

Feedback or experience from the case should be of interest to all industry-related practitioners and firms, no matter their venue and activity speciality. Industry-related and academic associations and publications are the best forums for dissemination of information like this.

Type B–I

The *Tucker Construction Company v Michigan Mutual Insurance Company* case

Jurisdiction: Florida (District Court of Appeal, Fifth District; on appeal from summary judgment of the Circuit Court, Seminole County; affirmed, December 15, 1982).

1 The Plaintiff
The Plaintiff was a construction company, presumed to have expertise in construction matters.

2 The Works
The works in question was a small restaurant of new construction in a suburban setting.

Tucker, the contractor, had hired a soils-testing firm, which took borings at the proposed site and made a recommendation that the foundations be placed on piles. Due to local zoning requirements, the building was constructed some 15 m (50 ft.) to the west of the test bores. No additional bores were made. The building was completed in October 1978 and handed over to the client.

3 Procurement Method
The procurement method was traditional, with the general contractor in a relationship of privity with the client and controlling all sub-contractors.

4 Insurance
The issue of contention in this case was property damage. The contractor had purchased a general liability policy for the project before commencing operations. The policy, from United States Fidelity and Guaranty Company (USF&G), provided cover until February 23, 1979, some four months after the building was completed. The contractor obtained a second policy, one for Completed Operations Liability (discussed under Type A above), on April 2, 1979 from Michigan Mutual Insurance Company. This latter was a comprehensive general liability policy, which Tucker claimed provided him protection.

The first policy type is generally understood within the US construction industry to provide cover for a contractor during the process of its construction or construction management activities to produce and deliver a project. Once the project is complete and handed over to the owner, a second type of policy with a separate premium is required. This second type is generally known as a *Completed Operations Policy*. It provides liability cover to the contractor from the time of the building's hand-over to the owner, subject to date of policy cover, for as long as the insured pays the premium required by the insurer.

Neither of these policy types purports to cover the liability assumed by a contractor through the terms and provisions of his contract to deliver a construction project as agreed. Comprehensive general liability policies provide cover to contractors for the usual tort liabilities arising from a failure to use due care resulting in personal injury or property damage. Although earlier court decisions fluctuated between finding for and against applicability of cover for property damage to the project (the product of the contractor's effort) itself, it was firmly established by the Florida Supreme Court in *Shelby Mut. Ins. Co. v LaMarche*[35] that cover for such damage was not provided by the normal construction of the policy phraseology:

> The majority view holds that the purpose of this comprehensive liability insurance cover is to provide protection for personal injury or for property damage caused by the completed product, but not for the replacement and repair of that product. ... Rather than cover and payment for building flaws or deficiencies, the policy instead covers damage caused by these flaws.[36]

5 Damage or Loss

This case involves neither personal injury nor damage to other property: the matter is economic loss and consequential damage traceable to the contractor's own or sub-contracted work.

The damage was excessive settlement of the building's main floor, which had to be replaced and prevented from further settlement.

The error was the placement of the floor slab directly on the suspect soil, without dedicated pilings or foundations. This happened because of lack of effective supervision, which supervision is presumed to include proper judgment on Tucker's part.

Pilings had been located under the foundation walls but not under the floor slab. Some 'flow settlement' had occurred during the period of USF&G-cover, and that insurer had paid for some repairs. Subsequent flow and settlement then occurred during the Michigan period of cover and that insurer asserted its rights to deny payment for repairs. The damage was discovered after completion, presumably by the client.

10 Process after Discovery

The contractor made a claim on the insurer, both in the instance falling under current operations, and presumably in the present instance, falling under completed operations. The contractor realised that he had a responsibility to remedy the defective floor. He mistakenly thought or hoped that his insurance would cover the situation.

The insurer was informed on presentation of the claim, but knew at the time it wrote the policy of the damages paid by USF&G for a similar defect.

Since the plaintiff was found to be at fault, the insurer of the plaintiff was also the insurer of the culprit. The case turned on interpretation of the policy wording.

11 Technical Reports

There is no indication in the appeal case report of any special technical ·report being made.

20 Resolution

The matter was resolved by litigation.

The insurer had filed a declaratory action asserting its rights and liabilities under the policy issued to Tucker. By summary judgment in the Seminole County Circuit Court, Robert B. McGregor, J., found for the insurer.

The insured appealed, with the case being tried in the District Court of Appeal of Florida, Fifth District. Cowart, J., presiding, affirmed the summary judgment with Orfinger, C.J. and Dauksch, J. concurring.

21 Costs

No information is available.

22 Information

There is no indication that any information was not available to all parties concerned.

23 Time

The claim was made by the insured some time after April 2, 1979, the date of the policy. Judgment for the insurer was affirmed by the Appeals Court on December 15, 1982.

24 Role of Experts

Not applicable.

25 Outcome

The contractor paid for his faults. The original insurer, USF&G paid for damage to the insured's product during operations resulting from defective workmanship. The second insurer, Michigan Mutual Insurance Company, providing cover during the completed period, subject to a claim for almost identical damage, did not pay. The disparity in treatment by the insurers can be attributed to different wordings of the policies with regard to work performed by or on behalf of the insured. The so called 'work performed' exclusion appeared in the 1973 Insurance Services Office's Comprehensive General Liability (CGL) form in wording that was frequently replaced after 1976 by the availability of a Broad Form CGL Endorsement that substantially narrowed the named exclusions. Many ambiguities of meaning were perceived by insurers, insureds and the courts, until various jurisdictions established firm positions on the matter such as Florida did in *LaMarche v Shelby Mut. Ins. Co.*[37], a Florida Supreme Court affirmation of a lower court ruling. The confusion over wording was compounded by early usage of the word *product* to designate a constructed building or project. This was a carry-over from manufacturing product liability insurance. It took until 1986 for this feature to be resolved by a version of the ISO CGL policy that clearly excepted 'real property' from the definition of 'your product.'[38] *LaMarche* and *Tucker* contributed to this process.

30 Feedback

Factual matters in this discussion are available to the public at large through the referenced law reports, particularly that of the title case.[39]

Type B–II

The *HOME OWNERS' WARRANTY CORPORATION and HOW Insurance Company, Appellants, v The HANOVER INSURANCE COMPANY, Appellee* case

Jurisdiction: Florida (District Court of Appeal of Florida, Third District; Sept. 4, 1996; on appeal from summary judgment of the Circuit Court, Dade County; affirmed, September 4, 1996).

1 The Plaintiff
As assignees of a condominium developer (Hoffman), Home Owners' Warranty insurance company (HOW) brought an action against the commercial general liability (CGL) insurer (Hanover) of the developer for its failure (breach of duty) to defend the developer against claims made by the condominium association for construction defects. In a summary judgment, Hanover was held not liable and HOW appealed with its principal, Home Owners' Warranty Corporation. The original related plaintiff was the Seville Place Condominium Association which had brought action against The Hoffman Group of Florida (a developer and builder), its associated companies and two construction companies.

2 The Works
The condominium was a new construction for two or more households.

3 Procurement Method
The procurement method with Hoffman as developer and builder was internal and tightly controlled.

4 Insurance
HOW was the original excess warranty insurer of Hoffman. Hanover was the original comprehensive general liability insurer of Hoffman for the project.

5 Damage or Loss
The original damage claimed by the condominium owners through the association was construction defects. Specific allegations included:

- Exterior walls designed and constructed in a manner which will not provide the Code-required wind-resistance.
- Exterior walls permit water intrusion.
- Interior finishes inadequate.
- Roofs permit water intrusion and not constructed in accordance with Code.
- Windows permit water intrusion.
- Exterior framing members deteriorating.
- Site drainage inadequate.
- Shower pans leak.
- Interior walls out of plumb.
- Roof trusses inadequately designed and constructed.[40]

The remedy claimed was compensation.

Compensation was consequential on correction of the damage listed above, identified prior to the initial claim against Hoffman.

The causes resided in workmanship, supervision and greed.

The damage was presumably discovered by the condominium owners.

10 Process after Discovery

The process was initiated by the property owners who presented their claim initially to the developer and builder.

The developer informed his liability insurer and warranty company.

The developer claimed that his liability insurer was obliged to cover damages and sought its support.

11. Technical report(s)

The facts as stated in the reference material indicate that a technical inspection and report were made and presumably paid for by the condominium association.

There is no indication of a report being commissioned by the defendant(s), but insurance companies normally conduct investigation of claims with internal staff personnel.

20 Resolution

The matter was resolved by litigation.

In the homeowners' original suit against Hoffman, Hoffman sought defence and indemnification by Hanover, its liability insurer. Hanover denied responsibility on the basis that cover for the allegations made was not incorporated in the liability policy. Hoffman, HOW and two others reached a settlement for approximately USD 1.6 million.

Hoffman then assigned its rights to HOW, who with its principal, Home Owners' Warranty Corporation, brought an action against Hanover for breach of its duty to defend and indemnify Hoffman. Summary judgment in favour of the insurer was entered by The Circuit Court, Dade County, Jon I. Gordon, J.

The assignees then appealed, with The District Court of Appeal, Third District, Cope, J. holding that the developer's liability for construction defects was not covered by the policy.

21 Costs

In the original settlement, the developer paid for its own defence as well as participating in the settlement damages.

There is no information at present on cost liability for the title case appeal and its basic suit.

22 Information

There is no indication that any party lacked access to information necessary to their arguments.

23 Time

There is no information available at present.

24 Role of experts

The technical reports (presumably made) were important, but the case turned on interpretation of the policy's exclusionary language; specifically that concerning cover for damage to the contractor's product arising from the product itself. Exception is provided for the exclusion if subcontractors have provided the damaged work or source of the damage. The Florida Supreme Court noted that 'an exclusion does not provide cover but limits cover.[41] Accordingly, the Appeals Court reasoned that, while the exception seemed to eliminate some sub-contractors from the exclusion, it did not create cover; therefore, the court denied a rehearing.

25 Outcome

The trial court decision was affirmed.

It was concluded that CGL cover was intended to provide an umbrella for instances of damage to persons or other property caused by property (the product) of the insured, not to pay for damage to or flaws in the property which arose from the property itself. 'Rather than cover and payment for building flaws and deficiencies, the policy instead covers damages *caused* by those flaws.'[42]

30 Feedback
Factual matters in this discussion are available to the public at large through the referenced law reports, particularly that of the title case.[43]

---

[1] Knocke, Jens 'Post-Construction Liability and Insurance' (1993) E&FN Spon, London, p.146.

[2] Miller, John B. & Fell, Mark C. <u>United States</u> in Knocke (footnote No 1) at p.354.

[3] Fla. Stat. § 768.76 (1993).

[4] Leiby, Larry R. 'Florida Construction Law Manual' Fourth Edition (1997) West Group, St Paul, § 15:13.

[5] *Putnam v Roudebush*, 352 So. 2d 908 (Fla. Dist..Ct. App. 2d Dist. 1977).

[6] Stein, Stephen G.M. <u>Warranties</u> in 'Construction Law' Rel.30 (1997) Matthew Bender & Co., New York, pp.18–1.

[7] *Raymond-Dravo-Langenfelder v Microdot, Inc.*, 425 F. Supp.614, 618 (D. Del. 1976) cited by Stein in <u>Warranties</u>, pp.18–2.

[8] Gonzales, V.M. <u>The Buyer's Specifications Exception to the Implied Warranty of Fitness for a Particular Purpose: Design or Performance?</u> in 'Legal Handbook for Architects, Engineers and Contractors' Dib, Albert, Ed., Vol 5, (1989) Clark Boardman, New York, as reprint from '61 Southern California Law Review' 237 (1987).

[9] Leiby, § 12:09.

[10] Leiby, § 12:09 and citing *Boscarino v Convenience Marine Products, Inc*, 817 F. Supp. 116 (S.D. Fla. 1993).

[11] Acret, 1995 Cumulative Supplement, p.170.

[12] *Miller v United States Steel Corp*, 902 F2d 573, 574 (7th Cir) rehg denied (Jul 12, 1990) cited by Acret.

[13] Leiby, § 15:01, citing twelve cases through most recent, *Casa Clara Condo Assn, Inc v Charley Toppino and Sons, Inc*, 620 So. 2d 1244 (Fla.1993).

[14] *Tropicana Pools, Inc v Boysen*, 296 So. 2d 104 (Fla. Dist. Ct. App. 1st Dist. 1974).

[15] *Wood-Hopkins Contracting Co v Masonary Contractors, Inc*, 235 So. 2d 548 (Fla. Dist. Ct. App. 1st Dist. 1970).

[16] Stein, Steven G.M. <u>Warranties</u> in 'Construction Law' Rel. 30 (1997) Matthew Bender & Co., New York, pp.18–6.

[17] *Flintcote Co v Brewer Co*, 221 So. 2d 784 (Fla. Dist. Ct. App. 3d Dist. 1969) cert. denied, 225 So. 2d 920 (Fla. 1969).

[18] Acret, James 'Construction Litigation Handbook' (1986) Shepard's/McGraw-Hill Inc., Colorado Springs, p.400.

[19] Acret, Ibid., p.392.

[20] Leiby, § 15:02 – citing *Caranna v Eades*, 466 So. 2d 259 (Fla. Dist. Ct. App. 2d Dist. 1985); *Hutchings v Harry*, 242 So. 2d 153 (Fla. Dist. Ct. App. 2d Dist. 1979).

[21] *STATE v GAF Corp.*, 747 P.2d 1326 (Kan 1987), 12 C.C. § 243, 'The Construction Contractor,' (1989) Federal Publications, Inc., New York.

[22] *Stephen v GAF Corp*, 242 Kan 152, 747 P2d 1326 (1987).

[23] Acret, p.274.

[24] explanatory note by author Turner, see following FN.

[25] Turner, S.C. 'Insurance Cover of Construction Disputes' (1992) Shepard's/McGraw-Hill, Colorado Springs, p.358.

[26] *Tucker Const Co v Michigan Mut Ins Co*, 423 So. 2d 525 (Fla. Dist. Ct. App. 5th Dist.) 1982) in Leiby, §20:14.

[27] *Southwest Forest Industries v Pole Bldgs, Inc*, 478 F2d 185 (9th Cir 1973).

[28] *Nixon v United States Fidelity & Guar Co*, 290 So. 2d 26 (Fla. 1973).

[29] *Hauenstein v St Paul Mercury Indem Co*, 242 Minn 354, 65 NW2d 122 (1954).

[30] *Missouri Terrazo Co v Iowa Natl Mut Ins Co*, 740 F2d 647 (8th Cir 1984) (claim insured floor sub-contractor for terrazzo floor that cracked, flaked, and settled: the exclusion of property damage to the products of the insureds was inapplicable, since damage was to the entire building, not just the floor).

[31] *Economy Lumber v Insurance Co of N Am*, 157 Cal App. 3d 641, 204 Cal Rptr 135 (1984).

[32] *Bowman Steel Corp v Lumberman's Mut Casualty Co*, 281 F2d 638 (3d Cir 1966)

[33] Wulfsberg, H.J. & Colvig, T.A. Insurance Aspects of the 1987 AIA Contract Documents in 'Design and Construction Contracts-New Forms, New Realities (1988) American Bar Association, Chicago, pp.294–5. Wulfsberg and Colvig refer to an excellent discussion of the 'work product' exclusions under the new CGL policies by Headrick and Wiezel, The New Commercial General Liability Forms – An Introduction and Critique, 36 Fed'n. Ins. & Corp. Coun. Q. 319, 356–68 (1986).

[34] *Chadwick v CSI, Ltd.*, 629 A.2d 820, 825 (N.H.1993), cited in Farnsworth, E. Allan 'Cases and Materials on Contracts' (1995) 5th ed., The Foundation Press, Inc, Westbury, p.394.

[35] 371 So. 2d 198 (Fla.2d DCA 1979), *affirmed*, 390 So. 2d 325 (Fla.1980).

[36] *LaMarche* at 326.

[37] 390 So.2d 325 (Fla.1980)

[38] Turner, S.C. 'Insurance Cover of Construction Disputes' (1992) Shepard's/McGraw-Hill, Colorado Springs, §§ 26.06, 27.01–27.05.

[39] *TUCKER CONSTRUCTION COMPANY, Appellant v MICHIGAN MUTUAL INSURANCE COMPANY, United States Fidelity and Guaranty Company, et al., appellees*, 423 So. 2d 525–529 (Fla.App.5 Dist. 1982).

[40] 683 So. 2d 527 (Fla.App. 3 Dist. 1996) footnote p.528.

[41] *LaMarche v Shelby Mutual Insurance Co.*, 390 So. 2d 325, 326 (Fla. 1980).

[42] *LaMarche* at 325.

[43] *HOME OWNERS WARRANTY CORPORATION and HOW Insurance Company, Appellants v The HANOVER INSURANCE COMPANY, Appellee*, 683 So. 2d 527–530 (Fla.App.3 Dist. 1996).

Uruguay

Type A

Aldo Lamorte Russomano

THE PURCHASER'S OPPORTUNITY FOR RECOVERY THROUGH LITIGATION OR OTHERWISE, FROM THE VENDOR

It is possible to advance three hypotheses:

(a) *That at the date of the sale the cladding was fissured.*

In this case, if it does not inspect or verify the object of the sale, the purchaser cannot seek to make the vendor responsible for manifest flaws (Civil Code (CC) Art1719). Here the principle of self-responsibility is applicable (an adoption of the French position). Opinions are divided concerning the possibility of claiming a reduction in price, *quanti minoris*. According to part of the doctrine, this is not incompatible with knowledge of manifest flaws. This view is not unanimously shared, and some consider that the purchaser lacks grounds for any action against the vendor.

(b) That at the date of the sale the defect in the material was not known (CC Art 1719).

By definition, this is a defect hidden at the time of the sale, thus allowing the purchaser to seek reparation. It is immaterial whether the vendor knew about the flaw, as he would have to make restitution even if he did not know. The purchaser can chose between revoking the contract with compensation for expenses incurred or claim a proportional reduction in price.

(c) *That at the date of the sale the vendor knew of the defect.*

If, in hypothesis (b), the purchaser can establish that the vendor did know about the defect in the material, he can claim full compensaticn instead of being limited to the expenses incurred under the contract.

Case Studies in Post-Construction Liability and Insurance, edited by Anthony Lavers. Published in 1999 by E & FN Spon, 11 New Fetter Lane, London EC4P 4EE, UK. ISBN: 0 419 24570 7

In all three situations, the right to action lapses six months after the acceptance of the thing sold.

THE PURCHASER'S OPPORTUNITY FOR RECOVERY THROUGH LITIGATION OR OTHERWISE, FROM THE VENDOR'S CONTRACTOR

Under hypothesis (a), the purchaser takes over the vendor's (i.e. client's or developer's) entitlement to claim for decennial liability. It is a case of private entitlement to subrogation. According to CC Art 1844, the contractor is passively put in the position of being liable for defects in the materials, even if these were supplied by the vendor. According to traditional doctrine, this principle refers to situations in which the stability or safety of the building is affected: the Construction Law of 1885 extends decennial liability to all defects.

The contractor is liable only for flaws that cannot be detected at the time of the reception and for the worsening of those which were apparent. In the case under scrutiny, liability would exist if the cracks were not detectable at the time of the reception or, if they were, if they subsequently became more extensive.

Under hypothesis (b), and according to a more modern view, which was used for the draft to reform the Law of Decennial Liability (with the participation of the authors and currently before Parliament), the solution is different: Since, in the case presented, the flaw only affects the appearance of the building, the severe provisions of CC Art 1844 are not applicable. Uruguayan construction law permits claims for all flaws, but the general duty of result does not apply to some of these: if a flaw does not threaten the safety of the building, then the contractor is not responsible if the material was supplied or required by the vendor. The contractor's liability is excluded if a party acts as an independent external cause. In such a case, the purchaser lacks grounds for action against the contractor.

THE PURCHASER'S OPPORTUNITY FOR RECOVERY, THROUGH LITIGATION OR OTHERWISE, FROM THE VENDOR'S ARCHITECT

The architect in charge of the design is not liable for the quality of the material, except in cases in which he supervises the works. In such a case, the solution is identical to that for the contractor (above). When both architect and contractor are sued, their concurrence in causing the damage must be analysed or both be liable *in solidum*, independently of the actions which they may take against each other.

THE INVOLVEMENT (IF ANY) OF INSURERS IN THE PURCHASER'S CLAIM FOR RECOVERY, INCLUDING THE METHOD OF RESOLVING THE MATTER

Construction insurance is not yet available in Uruguay. But, even if it were, the results would be the same, due to the lack of direct action against insurers; the insurers would undertake the defence of the architect or contractor and pay in case of an unfavourable ruling.

THE PROCEDURES BY WHICH THE ABOVE ARE RESOLVED, INCLUDING EXTRA-LEGAL, LEGAL AND INSURANCE PROCEDURES

Normally, the conflict would be resolved at a civil trial, through hearings in two instances and eventually (depending on the amount of the claim), where matters of law are in dispute, in the Supreme Court.

The parties may also agree in contract to settle disputes through arbitration. The arbitrators, through a procedure prescribed by law, hand down a verdict which puts an end to the controversy and is as binding as a court's ruling.

In Uruguay, no system of extra-judicial mediation has yet been implemented; what exists is binding conciliation before a magistrate as a necessary prerequisite to formulating a judicial claim, and afterwards intra-procedural conciliation presided over by the acting judge.

In the preliminary hearings (General Code of Procedures Art 341), the magistrate must endeavour to achieve conciliation between the parties.

ANY MECHANISMS EXISTING FOR UTILISING THE FEEDBACK OR EXPERIENCE FROM THE CASE TO A BODY OR BODIES WHICH MAY MAKE USE OF IT

No mechanism for feedback has yet been implemented. Since there are, at present, no insurance companies demanding such information, there have been no incentives to set up institutions compiling experiences and case studies by which professionals and institutions could learn from past mistakes.

Fortunately, the sectors involved seem to have realised the need for implementing such mechanisms.

Type B

Aldo Lamorte Russomano with Dora Szafir and Carlos Altoberro

Type B–I

The *Pando* case

Relevant to the Department of Canelones

1 The plaintiff

The plaintiffs were two Swedish nationals, resident in Uruguay because of a scholarship, who assigned the construction of their house in the Department of Canelones to an architect who was entrusted with the design and the direction of the works.

The *Banco Hipotecario del Uruguay* (Mortgage Bank) participated in the financing. By administrative law, this makes the Bank the beneficiary of any legal action concerning decennial liability. As such, this institution is entitled to claim damages in cases of construction defects or damage.

The plaintiff is, in civil law, the legal person who ordered the construction of a dwelling unit.

2 The Works

The project was a bungalow. The owners took out a mortgage with the *Banco Hipotecario del Uruguay,* which financed 85 per cent of the cost, including the architect's and the contractor's fees. The Bank's payments were staged in accordance with the progress of the works.

The clients engaged architect A to draw the plans and write the technical specifications for the project, as well as to undertake the technical direction of the works. Later on, they revoked the contract and engaged architect B, who took over when the works were done with the walls up to the level of the tie-beams.

The owners transferred to the *Banco Hipotecario* the right to claim compensation under the decennial liability of architect B as director of the works.

A year and a half later, and in mutual agreement with architect B, the owners put an end to his contract and engaged architect C for the remaining tasks; these were minor and did not prevent the owners from occupying the house.

Architect C took over the works, which had been executed in accordance with the specification drawn up at the beginning. Facing the risk of becoming responsible for architect's B actions, architect C also prepared a record of the inspections and technical checks carried out prior to his taking over, in which the existence of defects had been established.

Through its City Planning Division, the *Junta Local del Departamento de Canelones* (Municipal Council of Canelones), recorded, in writing, anomalies noted in the building, in its sanitary installations and in finishing works.

3 Procurement Method

The contractual method used was the traditional contract between owners commissioning works and an architect engaged for design and management.

Legally speaking, the person who acts as a contractor is the Client, the work force being engaged in his name, but the technically responsible person is the architect; in this case this person was also responsible for the design.

In Uruguay, all contractors must employ a person responsible for the technology used (an architect or an engineer). However, in owner-occupied construction, where the client-builder is a lay person, the one who assumes all functions, beyond purely formal aspects, is the architect.

The architect, in Uruguay, is a professional who may assume the one or several of the following roles:

(a) Designer and sole party responsible for graphic and written specifications;

(b) the project manager on the site: sole party responsible for correct and proper execution of the works, supervision of selection of adequate materials and technologies, in particular construction details during the course of works;

(c)   the responsible person in a contracting enterprise, i.e. the contractor's technical adviser and the person who assumes the responsibility for the works.

In medium-sized enterprises, it is customary that the same professional assumes roles (a) and (b), representing the Client's interests vis-à-vis the contractor, while in (c) it is the opposite.

In owner-occupier housing projects, the three roles may be assumed by the architect or engineer, given that the position of contractor is non-existent. It is usual that the architect contracts in the Client's name, but the architect is not an independent contractor since the works are executed as 'Owner's Direct Management' [The English term is 'direct labour': 'The employment of building tradesmen and labourers by the client or his agent (engineer or architect) directly, without the mediation of a contractor.' A Dictionary of Building, Second Edition, Penguin 1974].

4 Insurance
The only construction insurance available at the time was that for accidents on site, which covers the personnel contracted for the works.

There was no professional liability insurance, as this is not customary in the country.

5 Damage or Loss
The defective parts of the works were the sanitary installation, the walls (damp penetration), the finishing works in general (cracks in the rendering, loose tiles), uneven floors, the quality of the carpentry and timber and the base of the floors.

The defects appeared while the works were directed by architect B. At a later stage, the new professional, architect C, confirmed the existence of defects and reported them in the reception certificate.

The defects were due to the poor quality of the materials, non-qualified labour and faulty technical management.

The damage was detected by the plaintiffs, who requested a judicial inspection with the assistance of an expert appointed by the judge.
The clients sought compensation for the damage and harm caused, namely for material damage and emotional distress suffered [*daños materiales y morales*], indexed (for inflation) and with appropriate interest.

10 Process after Discovery

The compensation claimed was USD 50,000. The judge at first instance, however, understood that a final amount had not been established and ordered it to be determined at a later stage.

The claim was initiated by the clients, who, however, lacked the legitimacy for such action, as their right to act had been handed over to the *Banco Hipotecario*.

The claim was presented before the Civil Court of Justice of the Oriental Republic of Uruguay (*Justicia Civil de la República Oriental del Uruguay*) against architect B as manager when the flaws became manifest after termination of his contract.

No insurance company was involved.

11 Technical Reports

The plaintiffs commissioned, from an expert they trusted, a technical report, at their own initiative and expense.

Later, the magistrate (below), assisted by a technician appointed by himself, made a visual inspection of the works.

20 Resolution

Summary of the judicial proceedings: In the first instance, a payment of damages was requested for property and other losses emanating from the non-fulfilment of the contract between the plaintiffs and architect B.

The defendant was duly notified of the claim, and replied within the established period of thirty days, stating that he had handed over the works executed according to the technical specifications, and that the flaws and defects of the claims were not attributable to him. He also filed a counterclaim for professional fees and loss to his reputation ('moral damage').

As mentioned, the court inspected the works, and examined the construction defects and 'the bad quality of the materials', as the verdict later put it.

The preliminary hearing followed the procedures established in the *Código General del Proceso* (General Code of Procedure) Art 341:
(a) The parties confirmed their claims;
(b) conciliation attempts proved unsuccessful;
(c) the facts having been established, the parties pleaded orally and a new hearing was scheduled to issue the verdict and its motives.

After the ruling there was an appeal, see below.

21 Costs
The claim's initiators (the plaintiffs) paid the costs.

22 Information
The parties involved had access to all the studies and data related to the problem.

23 Time
The time spans for each stage of the process of resolution were one year in the first instance and one year in the second instance.

24 Role of Experts
The parties used their own experts, but the Uruguayan system does not admit such reports as proofs in the proceedings: only the court's expert's report is admissible; expertise cannot be replaced by other means.

The court appointed the professional assigned to do the technical report (above), and took into account the testimony of different parties involved in the construction of the works.

25 Outcome
The ruling in the first instance favoured the plaintiffs and required the defendant to compensate the plaintiff for the harm caused during the stage of execution; it also rejected the defendant's counterclaim.

The defendant filed an appeal. The plaintiffs were notified of this, and the proceedings were remitted to the corresponding Court of Appeal.

The writs having been received, the superior court revoked the first instance verdict: the plaintiffs' claim was rejected as they had accepted the works as being in conformity with the contract at the time of the revocation of the contract; in fact, the plaintiffs had (erroneously) understood that, by signing the reception certificate, they acknowledged that the premises were ready for occupation despite the flaws, while the magistrate held that 'conformity' meant that the works were accepted, with their flaws. Furthermore, the plaintiffs' claim lacked legitimacy, for they had forfeited the right to act on decennial liability in favour of the *Banco Hipotecario*.

It should be pointed out that the judge in the first instance, on the basis of the above mentioned lack of legitimacy, could and should have dismissed the claim at the beginning of the proceedings or in the preliminary hearing, thus avoiding an unnecessary procedure, given that the procedural system (General Procedure Code Arts 24.2 and 133.9) entitles him to do so.

Since the amount of the claim was not large enough for an appeal to the Supreme Court, the ruling in the second instance remained in force.

30 Feedback
Information on the case is open to the public and also via the legal journal *La Justicia Uruguay* (Uruguayan Justice).

Type B–II

The *Omega* Case

1 The Plaintiff
The plaintiff was a housing co-operative established solely for providing housing for each of its members, who are not owners but users. In this case there were 35 members.

The works were financed by a public institution specialising in housing with long-term loans, low interest rates and high subsidies.

2 The Works
The project was a residential complex, with units grouped in rows on one or two floors (duplex), of traditional construction with load-bearing walls of ceramic bricks.

The works started in December 1972/January 1973, and the hand-over took place in December 1980.

The duration of the construction period was eight years, compared with two years as an average for similar works.

The value of the works at the time of reception was some USD 170,000.

3 Procurement Method
The project was implemented through self-help building, i.e. through a system created under the umbrella of law 13728, the National Housing Plan of December 1968. The law appoints one public body for promotion and construction and another for financing and administration.

4 Insurance
The insurance covered liability for damage to third parties during the construction process. At the time of construction, the country lacked both property damage insurance and professional liability insurance (this type

of insurance may be offered in the near future, since the insurance market has been opened with the abolition of a monopoly law).

5 Damage or Loss

The affected part was the structure: cracks and caving in of such importance that they threatened ruin in the short and medium term.

The claim implied repair of the works by the person responsible for the damage (repairs *in natura*, rectification of the works) or, failing that, pecuniary indemnity for the same. In this case, the plaintiff preferred the second alternative. Given that the client had transferred its right of action to the financing institution (cf. Uruguay Type B-I), it was the latter who initiated the claim.

The site was part of a quarry. The deepest pit within the quarry, which represented an important portion of the total area, was used as a dump for oil and acid waste from automotive batteries. The pit was later filled with household waste. The latter, being organic in origin, decomposed with time, creating a black and smelly substance by lixiviation, which, under the pressure from the structures resting on the in-fill material, oozed to the surface, affecting three of the six rows of dwellings which were built here. Two of the units affected were one-storey dwellings, the others in two levels (duplex).

As a result of the decomposition process, the ground in the in-fill zone lost its load-bearing capacity and started to subside. This subsidence was uneven and, the units being linked, provoked not only fracture of the houses directly on the in-fill but also of those built on firm ground when they were dragged down.

Although the claim for defects originated from users who had been occupying the units for one or two years, some occupiers had pointed out that during the construction it was known that lixiviation occurred in some locations. However, the lack of the informants' technical competence and the absence of documentation concerning the process of construction made it impossible to corroborate those statements.

Several errors contributed to the situation:

(a) Lack of soil analyses and study of the subsoil.

(b) Use of inadequate foundations: stepped platforms of insufficient thickness (8 cm), which also precluded any movement; exaggerated length of the platforms which encompassed either five or six dwellings (40 metres); expansion joints which did not reach the foundations.

(c) Lack of a contracting firm which could have contributed the necessary experience to the execution of the works.

(d) A considerable proportion of the labour force was lacking the necessary qualifications.

(e) Inadequate supervision and technical direction.

The damage was, as mentioned, detected by the occupants of the dwelling units.

10 Process after Discovery

The claim was initiated by the occupants of the dwellings through the co-operative which owns the estate.

The claim was submitted to the public financing institution. According to its Charter, the rights of action against the architect-designer and manager of the works are transferred to it, since the works are the only collateral for the loan.

The Civil Code's Art 1844 determines the liability of the professional.

In this case, the claim was presented to the architect in person, though not through the court.

11 Technical reports

For details about the technical reports see 20(c)ff.

Since the occupants of the dwellings presented their claim to the financing institution, the co-operative never had to seek the advice of a legal expert who could have requested a report.

20 Resolution

The professional responsible for the works (the architect, designer and technical manager of the works) was insolvent, which was known in advance. An accord was reached between the affected parties: the owner (the co-operative) and the mortgage creditor (the financing institution) whose security was the property.

Summary of the procedure

(a) The co-operative presented the case to the financing institution, showing its technicians' report.

(b) The financing institution took various administrative actions to confirm the damage, but neither the origin of the damage nor the possibility of immediate reparation was analysed.

(c) The financing institution was advised both to analyse the causes of the problem before implementing solutions to repair it, and to determine the liability for possible claims. To this end, experts specialised in building pathology were brought in.

(d) First, a series of in-depth soil analyses were undertaken, and a preliminary approach made to determine the area occupied by the pit of the quarry, through extraction of the decomposing organic matter and mapping of the damage to the housing units.

(e) A proposal was made for underpinning by means of: (i) Cutting the foundation platform under the expansion joint; (ii) placing two prefabricated concrete or metallic beams under the platform; and (iii) positioning the beams to be supported by the head of piers outside the built area of the housing units.

(f) Due to the high cost (equivalent to three new housing units), this solution was rejected.

(g) New soil studies were carried out to determine the depth of the in-fill and the level of the firm stratum. Also, gauges were placed on cracks in the walls to study the progress of the phenomenon.

(h) A third series of soil studies was executed to determine the area with organic in-fill and establish exactly which of the affected houses would deteriorate further by the continued development of the phenomenon.

(i) Topographic mapping was undertaken to measure the subsidence and its evolution. Observation of the gauges (above) revealed that the phenomenon was still active and acute.

(j) Decomposing organic material was extracted and chemically analysed to determine its nature. From this, it was concluded that the decomposition process had not yet finished: the process of deterioration would continue, and the ground would not stabilise in the short term.

21 Costs
Since the process affected the collateral for the mortgage and the financing institution is a (public) mortgage institution (*Banco de la Vivienda*), the latter bore the costs, retaining the right to take subsequent legal action against the architect. However, no such action was commenced against the architect, since it was certain that no compensation was to be obtained due to the architect's insolvency.

22 Information
All parties had ample and free access to the information. Although the in-depth technical studies were carried out by the financial institution, it obtained from the owner (the co-operative) the commitment to engage the co-operative's own technical advisers – two civil engineers – to whom the building society periodically sent the studies and its own experts' technical conclusions.

The party responsible for the defect (the architect) showed no interest in the case, citing his retirement from professional practice.

23 Time
The process took approximately two years, during which the deterioration of the dwellings continued.

24 Role of experts
As indicated under 22, the mortgage institution and the co-operative had their own technical advisers.

25 Outcome
The parties arrived at the following agreement concerning the occupants of the dwellings affected with imminent or short term risk (six units):

(a) First, relocation of the occupants to units recently finished by the financing institute situated a few blocks away, where the occupants would be required to pay a moderate rent until their relocation;

(b) second, it was proposed that the financing institute provide either a loan (with an amount equivalent to the value of an unusable housing unit, with the same rate of interest and for the same period) for building new houses on neighbouring sites or a loan to buy dwellings on the open market or preferential treatment if they wished to acquire units in other housing complexes built by the financing institute.

(c) It was established that four units were in imminent danger of collapse with a further two soon to follow. One of the houses was buttressed and sealed off; the second was only buttressed (with the occupier refusing to move); a third, which was not in danger of collapse, was left vacant by its owner. Only one occupant accepted the solution of a temporary rental at a moderate rent.

(d) Although the solutions proposed above were accepted, the only units vacated were the two mentioned above, while the rest remained occupied.

(e) The Municipality of Montevideo and the Fire Department were informed of the risk and of the measures adopted by the financing institute. The latter also declined liability for subsequent developments.

(f) To this date the situation remains unchanged.

The experts advising the Client (the co-operative) were notified of all the actions as well as of the agreements, which were accepted.

References

Unofficial translations of the Civil Code

Article 1718
The vendor is liable for the defects or hidden defects [*vicios*] of the thing sold, whether real estate or chattel, if these render the thing unfit for the use for which it is intended or if it impairs this use in such a way that, had the buyer known of their existence, he would not have bought it or would not have paid the amount proffered. But the vendor is not responsible for patent defects or for defects which are visible, as he is not for those that are not visible but were known by the buyer, or could have been easily known by the buyer by means of his profession or trade.

Article 1719
The vendor must repair hidden defects, even if he did not know of their existence, when there is no provision to the contrary.

Article 1844
The architect and the contractor of a building are responsible for ten years if it [the building] suffers total or partial ruin due to a defect in the construction, a defect in the soil or bad quality of the materials, whether or not these have been supplied by the client, and in spite of any provision to the contrary.

The period during which the action can be brought is ten years from the time of the reception, but, once the action has been initiated after the manifestation of the defect, the time allowed is that of ordinary civil actions.

The disposition of the first paragraph should be understood unless a contrary proof is produced by the architect or the contractor ... .

Unofficial translation of the Law of July 8, 1855

Article 35
The producer [architect, contractor,...] of works is personally liable to the client according to the law for the period of time the law [the Civil Code] indicates for defects observed in the works, whether these be due to faulty direction of the works, to bad quality of the materials used, to workmanship, or to modifications or alterations introduced in the building without acceptance by the client and authorisation by the competent office, with written proof thereof on plans and specifications.

PART THREE

Analysis and Conclusions

It is, of course, largely impossible to generalise from the responses from the 19 countries received in the sense of discovering supra-national trends or patterns. Even between those countries which have a shared cultural or legal heritage, variations in practice were expected to be considerable and the content of the responses bears this out. However, these truisms do not lead to a counsel of despair; the fact that everything cannot be achieved did not mean that nothing should be attempted.

Accordingly the Editor tries in this Chapter to extract points of interest, especially by way of comparison and contrast between the responses.

Type A

The prospective respondents were asked to address five issues in their submissions and these will be treated in turn.

The Purchaser's opportunity for recovery, through litigation or otherwise from the vendor

It was not to be anticipated that respondents would rule out any possibility of recovery under any circumstances. Fraudulent concealment of known defects by the vendor would typically afford a right of damages or of rescission of the contract at the instance of the purchaser. Nevertheless, there were responses which were distinctly pessimistic about the chances of recovery in some jurisdictions. **England and Wales** starts from the maxim *caveat emptor* in contracts for the sale/purchase of real property (real estate), meaning in approximate translation 'let the buyer beware'. A properly drafted conveyance (the document effecting the transfer of ownership) would normally confirm this general effect. A similar presumption would apply in those jurisdictions which derive from English law. Recovery in these circumstances would thus be unlikely in **Hong Kong**. The Type A response for **Australia** states expressly that the maxim *caveat emptor* would apply there and that, because the development was com-

Case Studies in Post-Construction Liability and Insurance, edited by Anthony Lavers. Published in 1999 by E & FN Spon, 11 New Fetter Lane, London EC4P 4EE, UK. ISBN: 0 419 24570 7

mercial, none of the statutory warranties which might avail a residential purchaser would apply. **Singapore** might be the most optimistic common law jurisdiction for a purchaser pursuing a vendor, given the developer's obligations under statutory regulations to use a standard form Sale and Purchase Agreement expressly binding him to erect the building unit 'in a good and workmanlike manner.'

This is not a peculiarity of the common law systems. Other respondents expressed serious doubts about the purchaser's chances of recourse against the vendor. A key factor in this was often the purchaser (Bank) failing to inspect the property. In **Germany** the fault would need to diminish the value of the building or reduce its fitness for purpose. The hairline cracks would not reduce its fitness for purpose as required. It is also doubted whether under the law of **Germany** aesthetic appearance would be considered a warranted quality. Inspection would be usual and failure to inspect might constitute 'gross negligence' on the part of the Bank. In **Sweden**, also, it is said the purchaser would have a responsibility to conduct a thorough examination and could not turn against the purchaser in respect of defects which could have been discovered during that examination. In **Japan** the purchaser would not be able to categorise the surface cracking as a 'concealed defect' for the purposes of claiming under a warranty of fault in the Civil Law. The Bank and the purchaser would have had to perform an inspection without delay after the acceptance of the building, giving the vendor immediate notice of any defect, if the purchaser were to be able to avoid the contract or claim damages under the commercial law. In **Kuwait** the purchase-sale contract could in theory contain a warranty upon which the purchaser could rely, but usually the expectation would be that the purchaser would inspect and accept the premises as they are. Failure to prove that the vendor misled the purchaser or was guilty of some other kind of fraud would result in a court dismissing the purchaser's claim.

The failure to inspect casts doubt upon the purchaser's rights of action against the vendor even in those jurisdictions where the opportunities for such action clearly existed. While the Civil Code of **Canada, Quebec** does not require inspection by an expert, it adopts the 'reasonable person' standard for discovery of defects for limitation purposes. Furthermore, there is a question as to whether the defect was 'hidden' for the purposes of the Civil Code. The Bank may not have acted as 'prudently and diligently' as the law would require from a 'professional' (i.e. expert) purchaser.

The legal system of **Uruguay**, like just about every other, will permit the purchaser to recover from a vendor who knew of the defect, and under the Civil Code the purchaser could get rescission of the contract or a pro-portional reduction in price if the defect is genuinely unknown to both parties at the date of sale. However, under the Civil Code of **Uruguay**, the purchaser who does not inspect or verify the condition of the subject mat-ter of the contract (the building) cannot seek to make the vendor responsi-ble for manifest flaws. The principle of 'self-responsibility' is explained as an adaptation from French law.

In **France** itself, the nature of the damage, which is aesthetic, would take the case outside the province of producers' liability under the well-known Article 1792 of the Civil Code. This would force the purchaser to rely upon the general law rather than the decennial liability provisions which are a salient feature of the system in **France**. The vendor would be more attractive as a target for a claim by the purchaser than the producers, because it is the only party with whom the purchaser has a direct contrac-tual relationship.

Of those systems where the purchaser has clear opportunities for action against the vendor, several deny the existence of that need to inspect found in those discussed above. In **Argentina,** the purchaser can sue the vendor even though the latter acted in good faith and in ignorance of the defects. Inspection by a purchaser is not obligatory and the Bank would not be negligent in failing to inspect. It is emphasised that in **Denmark** it is of no importance that the purchaser did not inspect the building which would have revealed the cracks, because the purchaser had no obligation to do so. The vendor's conduct is equally immaterial: the vendor is likely to have to pay damages whether he was negligent or not. It is worthy of note that in **Denmark** the purchaser would have to act promptly to repair the building as soon as the court-appointed experts had reported; any costs at-tributable to delay in repair would be irrecoverable from the vendor.

The respondents for the **United States** quote a standard text in the Florida jurisdiction as saying that 'if the purchaser has a reasonable op-portunity to discover a defect prior to sale and fails to exercise that op-portunity, then failure to discover the defect can be a defence to a breach of warranty action', and they add the opinion that the Bank's failure to in-spect has left it in a weak position to claim against the developer-vendor. Generally, the purchaser could only look to an express provision in the contract of sale as against the vendor, although a transfer of the vendor's rights under a manufacturer's warranty of the product under the Uniform Commercial Code is also discussed.

An action against the vendor could be possible in **Switzerland**, al-though it would not cover damages for economic loss, the remedy poten-tially available being limited to rescission of the contract or compensation for diminution in value of the building. The purchaser's action would exist independently of any culpability on the part of the developer vendor.

The prospects for recovery by the purchaser against the vendor in **China** would depend upon express provision in the contract and upon the (two year) limitation period for commencement of the action not being exceeded, under the provisions of the General Principles of Civil Law (or Civil Code). While this gives a different emphasis from that in **Hong Kong**, which belongs, of course, to the common-law tradition, the frequency of such express provision as would protect the Bank in the Type A scenario may not be any greater than in **Hong Kong**, where express provision is also possible, though rarely used.

Although in some systems, for example, **Canada, Quebec**, it has been doubted whether the defect in question was hidden, under the law of **Chile**, largely as a result of interpretation in case law, it is enough that a 'seed' of the latent defect existed at the time of sale. That would offer the purchaser the prospect of a claim for a 'redhibitory action', claiming rescission of the contract. The alternative would be a *quanti minoris* action for diminution in value; the contractual remedies potentially available in **Chile** can thus be compared, in broad terms, with those available in **Switzerland**. In **Chile**, the purchaser can also consider bringing an extra-contractual action, based on negligence or fraud, within a four year prescription period. This possibility is mentioned in a number of other Type A responses, but is dependent on some fault on the part of the vendor being shown, which is by no means clear-cut from the facts supplied.

The Civil Code of **Spain** places emphasis upon the question of whether the defects in a building are hidden vices (*vicios ocultos*); the vendor can be liable to the purchaser for these even in the event of its ignorance of them.

Probably the most forcefully expressed answer in favour of the purchaser's ability to recover from the vendor is that of **Norway**, where the Sale of Real Estate Act 1992 governs the situation. If the cracking was not obvious, but could only be discovered by a careful inspection, it is quite clear that the lack of inspection does not defeat the purchaser's claim. Although negligence by the purchaser would in principle be capable of reducing the damages payable, the response for **Norway** is that the cracking made the building unsuitable for its intended purpose, thus constituting a breach of contract by the vendor.

The purchaser's opportunity for recovery, through litigation or otherwise, from the vendor's contractor.

The purchaser's opportunity for recovery, through litigation or otherwise, from the vendor's architect.

It is convenient to treat together these two possible avenues of recovery by the Bank against producers employed by the vendor. Although different issues are likely to be involved between the two, because of the architect's status as professional adviser and because the architect protested at using a product of which he had no knowledge, neither the architect nor the contractor has any direct contractual relationship with the Bank. Whereas consideration of the Bank's ability to recover from the vendor was largely grounded in the contract of sale or some form of contractual warranty, attention now turns to the availability of recourse outside of contract against producers who are 'third parties' in the sense that they have no direct contractual relationship with the purchaser.

Overall, the responses for the majority of countries submitted were negative about the prospects of recovery against third party producers. Nevertheless, several interesting features emerge from the responses on these points.

First, as between the countries in which recovery against the contractor or architect was said to be unlikely, there is a difference between those systems where the barrier is one of principle and those where the facts given do not satisfy the requirements for such an action. Second, unlike the issue of action against the vendor, there is no consistency between the common law jurisdictions; on the contrary there is a sharp division of position.

Finally, there are notable exceptions to the majority trend against third party producer liability where there are said to be good prospects for recovery against the contractor and architect and where the basis of potential liability will not be a simple tort claim.

The common law systems, in an area of law where there has traditionally been considerable consensus, reveal recent but fundamental disagreement.

Following major appellate court decisions in the 1980s, in **England and Wales**, it has been very difficult for tenants or remote purchasers (i.e. successors in title to the original developer) to sue third party producers, such as contractors and architects, employed by the original developer. This is not because they cannot owe a duty of care in tort, but because that duty is basically limited to liability for injury to persons or damage to their property, to the exclusion of economic or consequential loss, even the cost of making good a defective building. The damage in the Type A case study is not of the kind recoverable. **Hong Kong** follows this position exactly; recovery of economic loss by the Bank against the third party producers is not possible.

However, **Australia** has adopted a sharply different position, in common with New Zealand and Canada (but see **Canada, Quebec** below). The response for **Australia** cites the important 1995 decision of *Bryan v Maloney* which departed from the late 1980s case law in **England and Wales** referred to above. That decision preserved the possibility in **Australia** of actions against negligent contractors or architects, although their failure to achieve the required standard must be proved by the plaintiff, in this case the Bank. Even in the event of the action being taken against the contractor, the limited aesthetic nature of the defect will limit entitlement to recovery; a Court decision from the State of Victoria on similar facts to the Type A scenario held that the owner was not entitled to a 'Rolls Royce' standard of rectification through replacement of the façade. **Singapore** has taken a similar position to that of **Australia** and *Bryan v Maloney* was cited with approval in *RSP Architects v Ocean Front*, the equivalent Court of Appeal decision in **Singapore**, in preference to the case law from **England and Wales**. However, while a right of action in tort for economic loss is possible in **Singapore**, on these facts it would not be likely to succeed. Certainly, the contractor was not negligent, having performed his duties and so could not be liable. The architect had also given evidence of reasonable care and had protested about the risk of the use of the grc product.

The **United States** position resembles that of **England and Wales** and **Hong Kong** in one important respect; the respondents quote a leading text as follows: 'Damages of economic loss are not recoverable under a negligence theory absence of privity of contract.' The prospect of an action against the architect in (the Florida jurisdiction of) the **United States** is held out by the fact that 'Architects in general have a duty of care responsibility as licensed professionals' but the conclusion on the premise that the architect had advised against the use of tiles without specific knowledge of the research into their deficiencies was that the owner was in the position of having bought what he asked for; the purchaser would not be placed in a better position.

The civil law jurisdictions, although the reasoning is usually quite different, mainly reach similar overall conclusions on the problems of proceeding against third party producers .

Perhaps the clearest denial of liability on the part of the producers is to be found in the response from **Spain**. The Civil Code provisions on decennial liability would not apply to this case because the defect in question would not come within the limits of 'ruin', defined by the jurisprudence of **Spain** as pertaining to 'unfunctionality of the building'. The law relating to liability for defective products could not be applied either. Even if such an action could be contemplated, the 'state of the art' defence would be available to the architect.

The response from **Denmark** is also very forthright in rejecting liability by both the contractor and the architect. The contractor under the law of **Denmark** has no 'duty to warn' of deficiencies in the design or specification if the likelihood of damage is obvious. The position of the contractor is simply that he performed his work according to specifications and has no responsibility for the purchaser's loss. While an action against the architect in tort could be contemplated in theory in **Denmark**, the architect's duty could not be greater than that to its client; the error in question was not 'an obvious fault' and the client had chosen to bear the risk of using an unknown new product.

In **Kuwait** the contractor could not be liable to the Bank under the Civil Law unless the contractor had been aware of the problem with the cladding panels, or had not complied with the specification. There appears to be some possibility of liability on the part of the architect for design or supervision, if the latter was also undertaken.

Japan largely rules out the possibility of recovery against the contractor in the absence of a contractual relationship. The chance of recovery of compensation under the Civil Law would be dependent upon the contractor having committed an illegal act of some kind. In **Japan** there would be more chance of an action against the architect because of the high onus of responsibility placed upon him to take care. The architect's concession to the vendor's wishes could be the key issue in **Japan**.

In the response from **Norway** it is clear that the purchaser will not be able to seek recovery from the contractor. This is not because of the absence of contract; in principle a subsequent owner could sue a producer employed by the vendor. However, the contractor is not in breach of contract, and this is decisive. The same theoretical possibility would exist in **Norway** for action against the architect, but the response places great weight on the client having insisted on taking the development (or state of the art) risk, so that the architect has not acted negligently. The architect's liability would be limited to a maximum figure of NOK 1.5 million.

Sweden also has some modification of the doctrine of privity of contract. The basic rule would be that no recovery for pure economic loss, as in the Type A scenario, would be possible in the absence of a contractual relationship between purchaser and producers.

In **China** the rights of the purchaser would be strictly limited under the Civil Code because the Bank was not a party to the contracts with the contractor and architect. The contractor could be liable in theory under the principle of product liability under the Civil Code, but the People's Courts are enjoined to 'distinguish right from wrong'; some element of knowledge or involvement by the contractor in the selection of the product would probably be necessary to establish 'wrong'. The response for **China** envisages that the architect would not be liable under the narrow concept of negligence, although joint liability with the vendor for any infringement of the rights of the purchaser would be possible.

The **Uruguay** response considers that the purchaser may, under the Civil Code, be able to take over the vendor's entitlement against the contractor for decennial liability purposes. Whereas decennial liability had previously referred only to defects affecting stability or safety, the Construction Law extended decennial liability to all construction defects. However, under the modern law of **Uruguay,** which has been the subject of reform, the contractor will not be liable under a general duty of result for a flaw which does not threaten safety or stability. The architect's position in **Uruguay** is described as identical with that of the contractor; if they are both sued, their liability will be considered *in solidum.*

The remaining jurisdictions offer some prospect of action against the contractor and/or architect, although this ranges from a strong to a somewhat theoretical possibility.

In **Switzerland**, the contractor could be responsible and the opportunity to recover would not be dependent upon negligence being proved. The architect could only be liable if proved to be at fault, although the burden would be upon him to prove that he is not, and on the facts of the Type A scenario that burden may defeat the architect. This is so particularly because the architect did not investigate the product so as to enable him to give proper warning.

France reports that the kind of damage in the Type A scenario will be interpreted in the French courts as outside the producers' or Decennial Liability provisions of the Civil Code, because it is purely aesthetic. The Bank's ability to turn directly against the producers will therefore depend on the general law (the equivalent of tort or extra-contractual liability), where, once again, as in so many of the countries previously considered, fault by the producer would be necessary. It is well known that one of the great attractions of the insurance provision in **France** is that it is not dependant upon proof of fault, but once a claim falls outside that 'envelope', an error must be shown. If, as is likely, the Bank turned against all the producers, the court would allocate compensation in proportion to the degree of fault which it found.

In **Germany**, the ability of the purchaser to sue the producer will depend upon contract, which distinguishes it from most of the countries considered in this section. This is because the vendor's warranty rights can be assigned. However, this would depend upon such assignment being made express as between vendor and purchaser. This is commonly done in **Germany,** but the respondents have found nothing in the facts to support such a claim. This position applies equally to contractor and architect.

Of the remaining jurisdictions, **Argentina** appears to offer the clearest opportunity for the purchaser to recover. The contractor could be responsible under the Civil Code if it has violated norms of professional/trade responsibility, particularly if it has not employed reliable personnel. The architect may incur liability under his design responsibility.

The response from **Chile** is more complex. As with most other jurisdictions, extra-contractual liability, which involves a limitation period of four years, will involve proving fault, either wilful misconduct or negligence. For the contractor, this would necessitate proving that it was involved in the choice of materials and should have ensured their quality. The respondents for **Chile** conclude that the architect's warning to the developer should release him from a finding of fault. The other possibility, an interesting one, is the potential liability of the producers under the General Law on Urbanism and Construction, which provides for liability on the part of contractor or architect as a person who has undertaken work involving risk. This is, in effect, a duty of result, although the respondents refer to defences.

The possibility of using a manufacturer's fitness for purpose duty to consumers under the Civil Code is canvassed in the response from **Canada, Quebec** as a means of obtaining redress against the contractor. The difficulty is that this provision of the Code has not yet been interpreted by the Courts as applying to real property (real estate). The Civil Code of **Canada, Quebec** also provides for the transmission of contractual rights, such as warranties of the quality of the property, to successors in title, such as the purchaser. This is a statutory means of producing the effect achieved by express contractual provision in **Germany** (and other jurisdictions).

The contractor's duty in **Canada, Quebec** is virtually a duty of result, at least insofar as concerns major defects. The respondent notes that the client's interference, as in the Type A scenario, might provide the contractor with a defence, but points out that development risks usually fall upon the contractor. The position of the architect is finely balanced because his liability will depend upon the standard of his professional conduct; the response from **Canada, Quebec** weighs up the factors in favour of purchaser and architect respectively, referring to Quebec case law suggesting that the architect may have breached its duty to investigate the quality of materials to be used.

The involvement (if any) of insurers in the purchaser's claim for recovery, including the method of resolving the matter.

Sharp divisions may be observed between what can be described as three groupings of countries and virtually every country for which a response has been received can be placed in one of the three. The first category is of those countries which have sophisticated insurance arrangements for the building and/or its producers. The second category insurance is concentrated in the professional liability field, so that the architect is the only party likely to be insured. In the third category, insurance arrangements are either absent or limited to inapplicable contractors' policies. However, care must be taken in supposing that the countries with established schemes are automatically better off in all situations. In the Case Study A scenario, the nature of the building, i.e. non-residential, and the nature of the damage, i.e. aesthetic, are such that the injured party could be little or no better off than in countries with more primitive arrangements.

France is, of course, the country with the best known and most studied system based on *dommage-ouvrage* insurance of the building for a period of ten years, and it offers a good route of recourse for purchasers and other successors in title to the original owner, covering the cost of repair of the structural parts of the building without the need to prove fault by the producers. The *dommage-ouvrage* insurers can then recover against the providers of the mandatory liability insurance. However, in **France** this case would not come within the decennial liability provisions of the Civil Code, being merely aesthetic damage. The developer is unlikely to have sought any insurance beyond that, which is compulsory. Therefore the developer will be uninsured as regards this risk. The contractor must have the obligatory decennial cover as a producer and there is a chance that he will have insured against post-reception risks beyond the decennial risks (i.e. beyond structural failure). The architect must by law carry insurance against all the consequences of his actions and errors, and so may be a more eligible source of recovery, subject to the vital question of liability discussed above.

Sweden reports on an insurance policy known as Insurance Against Latent Defects with a ten-year duration, but, again, cosmetic defects fall outside its scope. Beyond that, an insurer would not expect to cover a defect which was built in and in existence at the time of purchase.

Australia is split into states with different legal and insurance regimes, but the system in Victoria has influenced other states. It offers one of the best opportunities of recovery for a purchaser of any of the countries covered in this work. Although the developer will not normally carry insurance, both the contractor and the architect must have mandatory cover for a period of ten years from the issue of the Certificate of Occupation following completion of the works. The contractor's insurance is to cover structural defects, and the aesthetic damage in question will probably fall outside it, but the architect's insurance is much wider, and an omission to carry out research would be covered, up to a limit of AUD 1,000,000.

Switzerland, as might be anticipated from the home of some of the world's major insurers, typically has some or all of its producers covered. These could include a policy to cover the developer's liability, albeit with limited cover, a policy covering the contractor's general business liabilities, which would not assist in this case, and professional liability insurance of the architect, which could, subject, of course, to liability being established.

Since 1997, in **Norway** a German insurer has offered a kind of building defects liability insurance covering defects in construction causing loss to the owner and successors in title during a ten-year period. This would not cover the risk of a product chosen by the developer, but if the defect was the responsibility of the architect or contractor, this insurance would pay and then seek recourse against the responsible producer or its insurer. Apart from this insurance product, in **Norway** only the architect's professional liability insurance could be relevant to a claim of this sort (but see liability discussion above), since the contractor's insurance, of the all-risk project type, would not cover it.

Although this kind of building defects liability insurance is available in **England and Wales**, it is not in widespread use and the English response concentrates on the professional liability insurance of the architect. Latent defects insurance has not been used significantly in **Hong Kong**, which, like **England and Wales**, would focus upon the architect's professional indemnity insurance. The building insurance which would be carried in **Hong Kong** would not govern this kind of damage.

Owners would not carry the kind of insurance needed to guard against this kind of damage in **Denmark** or in **Germany**, where there is no tradition of building defects insurance, the feeling being generally hostile to a system which could protect producers from the consequences of poor quality work, funded by the premiums of those with higher standards. The bond system in **Germany** serves to ensure that contractual obligations are fulfilled and the rights under a bond or warranty are transferable by express provision.

In **Chile** it appears that types of guarantee insurance have been developed which fulfil the same purpose as performance bonds. Guarantee policies can be issued for hidden defects but there is no widespread post-construction insurance in **Chile**; such cover is limited to one to two years, and premium levels are high. Most insurance in **Chile** is limited to the period of the work. An interesting development is that a specific construction company in **Chile,** which has defined standards of quality, has offered a three year guarantee to purchasers, which it has backed with insurance cover.

The contractors' insurance routinely obtained in the **United States** would not be expected to cover this type of risk. The same is true of **Kuwait** where contractors' cover is normally limited to Contractors' All Risk. The architect in **Kuwait** would be insured to cover liability for loss or damage arising from professional negligence, because professional indemnity insurance is required by law. In **Japan** there is currently no property insurance or warranty insurance capable of insuring this kind of damage. Developers carry liability insurance, but not for building defects like this. Architects' liability insurance would be relevant, although a major difference in practice from Western countries should be noted. Whereas Western insurers habitually become involved in disputes, controlling and conducting litigation, in **Japan** an architect's liability insurer would at most offer advice to a negligent architect. The response from **Spain** identifies the advantages to the purchaser in pursuing a claim if the vendor was insured, but the probability of this covering the situation is rated as 'low', because of the nature of the defect, which does not affect stability or performance.

The architect would be the only party who would be insured against the kind of risk in the Type A scenario in **Canada, Quebec**. The developer's insurance, like that in most North American jurisdictions, would typically exclude the cost of making good faulty materials, workmanship or design. The contractor's insurance would not cover this kind of liability, although it may be noted that in **Canada, Quebec,** as in **Chile** and many other jurisdictions, performance bonds are used to secure proper contractual performance, and these may be extended for one year (a typical duration) after completion to cover the usual guarantee period. The architect's insur-

ance would therefore make him a prime target for those suffering loss in
Canada, Quebec.

The South American jurisdictions for which responses were received
demonstrate little use or availability of insurance of this kind of risk.
Some recent developments are reported from **Argentina**, where insurers
have begun tentatively offering liability and professional indemnity cover,
although premiums are high and renewal uncertain, and cover limited to
ARS 50,000. The position is less advanced in **Uruguay** where professional liability insurance is not well established. In **Singapore**, perhaps
more surprisingly, given the stringent nature of liability, or perhaps resulting from the insurers' nervousness of this, producers do not habitually
carry insurance. Developers do not insure, contractors only carry Contractors' All Risk cover and professional indemnity cover of consultants
such as architects is far from complete. The respondents for **China** can-
vass the possibility of the grc tile manufacturer carrying product liability
insurance but nothing in the Civil Code suggests the involvement of an insurer in **China** in a post-construction liability dispute, nor does the Procedural Code provide for an insurer's involvement in litigation.

**The procedures by which the above are resolved, including extra-legal
and insurance procedures.**

**Any mechanism existing for utilising the feedback or experience from
the case to a body or bodies which may make use of it.**

The ways of resolving disputes reveal commendable ingenuity and diversity as between different legal systems. There is a widespread, although
not universal, desire to avoid litigation, either because a developed industry has negative perceptions of the cost and other harmful effects or because of cultural aversion to confrontation. Readers wishing to go beyond
the limited scope of this section of the analysis are referred to Dispute
Resolution and Conflict Management in Construction (eds P. Fenn, M.
O'Shea and E. Davies) published by E&FN Spon London, 1998 (ISBN 0
419 23700 3), which arose out of a specialised international study by the
former TG15, now CIB Working Commission 103.

Amongst the most sophisticated systems for dispute resolution, it is
common to find two features, which can be related, namely specialist tribunals set up to deal with construction or even post-construction cases and
alternative mechanisms (sometimes collectively known as Alternative

Dispute Resolution or ADR). These take their place among a whole range of procedural options.

In the State of Victoria in **Australia,** statute has intervened in the shape of the Building Act 1993, both to try to improve the fairness of liability litigation by abolishing joint and several liability and to order the parties, where appropriate, to engage in mediation, which is now common. To these must be added reference by the Court to a specialist construction referee, arbitration where agreed by the parties and the traditional form of litigation. **Australia** generally, and the State of Victoria in particular, are well-endowed with methods for resolving post-construction disputes. Similar levels of sophistication can be found in **Canada, Quebec**. Whereas a non-residential case like type A could not be handled under the Quebec New Homes Warranty Program, which would largely avoid the risk of litigation, there are several other possibilities. The preference will be for an out-of-court settlement and the parties may try voluntary mediation, if necessary with the courts' assistance or that of the Quebec Bar Association. Arbitration may well have been agreed in the contracts with the producers, although probably not in the sale contract, as an alternative to the court system. **Sweden** has a special tribunal for resolving disputes involving architects or engineers, in addition to its preferred arbitration or judicial tribunals.

In **England and Wales**, where mediation is also becoming much more common in the construction industry (see Case Study B–II), the High Court has a specialist construction court. Formerly the Official Referees, since October 1998, this body has been known as the Technology and Construction Court. Arbitration has been re-vitalised by the Arbitration Act 1996.

As was the case with **Canada, Quebec**, in **Germany** the Court may assist the parties in trying to reach a compromise to avoid civil litigation or arbitration, the traditional procedure.

A number of responses emphasised the frequency of settlement. In **Switzerland,** the presence of insurers in a dispute is seen as a force for settlement. The rules of civil litigation can be complex, as they vary from canton to canton (administrative government units). Negotiations commonly end in settlement in **Norway,** where arbitration is the most commonly preferred alternative to litigation, although the legal issues raised by the type A scenario might lend themselves to litigation rather than to any kind of alternative. Negotiation is also a very common way of settling such matters in **Chile**, although arbitration clauses in construction and construction-related contracts are said to be common. The expectation in **Spain** would also be that both the sale-purchase contract and the contract for the supply of the tiles would contain arbitration clauses. If litigation did result, insurers could in principle be involved if the subject matter was insured.

Uruguay has a practice of reference to arbitration pursuant to an arbitration clause, but no tradition of mediation or other ADR techniques, so arbitration is the principal alternative to civil litigation.

The parties to a dispute in **Argentina** can attempt extra-judicial meetings to try to clarify the matter with the help of technical and legal experts and arbitrators, but otherwise the court is the principal forum for hearing the matter.

Some Asian jurisdictions have a strong preference for settlement as culturally more acceptable than confrontation. In **Hong Kong,** this can be promoted by techniques such as mediation; in addition, arbitration is common in the construction context. The court system is based on that of **England and Wales.** If the facts of the Type A scenario occurred in **Japan**, the Bank would expect to commence negotiations with the vendor. The vendor would seek to involve the contractor and architect in the negotiations, seeking contributions from them as a way of avoiding a dispute. If no agreement is reached, **Japan** has two legal processes, namely conciliation and litigation, to produce a solution. Insurers are rarely involved at advanced stages of disputes. **Singapore** has a tradition of arbitration as well as litigation and reference can be made, for residential properties, to the Conciliation Scheme organised by the Real Estate Developers Association of Singapore (REDAS). **Kuwait** also would expect a resolution through negotiation between the Bank (the purchaser) and the developer, failing which litigation will be essential to obtain a verdict of the court which will be enforceable.

In **China**, there is a place for arbitration, but it is limited to cases involving 'foreign economic interests'. Disputes between citizens of **China** or companies registered there will be resolved in the People's Courts under the Procedural Code.

Finally, the role of independent experts should be acknowledged. (See also the analysis of the Type B Case Studies below). In **France**, a tribunal of such experts will resolve any disputes between the *dommage-ouvrage* insurers and those of the producers, who insure their decennial liability. In **Denmark** the court appoints independent experts to report before the parties embark upon litigation or arbitration.

As with dispute resolution, there is inevitably something of a hierarchy of sophistication when it comes to feedback. A number of responses gave a nil response to the question on feedback, either because there is none or because the respondents chose not to treat it. This applies to **Chile,** to

Switzerland, to **Kuwait**, to **Canada, (Quebec)** and to **China,** where the Code makes no relevant reference. The absence of insurers who would be natural providers of feedback, is blamed in the responses of **Argentina, Uruguay** and the **United States.** The response for **England and Wales** does not treat feedback, although insurers and technical bodies could probably provide some examples of the process. No feedback generally is also noted for **Sweden** and for **Japan.**

In a number of the remaining jurisdictions, feedback is limited to law reports which will not provide systematic or expert feedback. This applies to **Spain** where it can be expected that case law, although not arbitration awards will be published and to **Hong Kong**, where feedback is limited to the law reports. In **Germany,** feedback comprises a combination of professional journals and published official law reports.

The final category of respondent countries have better standards of feedback in one way or another. **Norway** has no institutionalised feedback, but the small size of the industry ensures that leading cases are well known. In particular, bad products, like the grc panelling in the Type A case, will become well-known quickly. In **Denmark,** not only are the Building and Construction Arbitration Court's decisions on legal findings published, but the decision can refer to settlement proposals before litigation which can also be published. **Singapore** also reports decisions of the courts in the official law reports, but in addition, feedback goes to government and other official institutions, such as the Building Control Department and official bodies such as the Real Estate Developers Association of Singapore and the Singapore Institute of Architects.

Australia has one of the most sophisticated feedback systems, with reporting to the Building Disputes Practitioner Society and the Building Codes Board, as well as the readers of the official law reports.

The system of SYCODÉS found in **France** is a model for other nations to copy in that the reporting of claims events is well-organised, informative and capable of analysis. **France** has much to teach other jurisdictions in the area of feedback.

Type B Case Studies

The Type B Case Studies are in most respects less eligible for analysis than the Type A responses, because they are, by definition, more disparate, making comparison difficult. They should in any event, having been selected as illustrative of procedures for handling post-construction liability issues, tell their own story.

However, it was considered worthwhile to add some remarks upon general themes with reference to certain of the Type B Case Studies. No disrespect is intended to those respondents whose Case Studies are not referred to specifically; the selection is purely to suit this writer's purposes and there is, of course, merit in producing a submission which requires no further comment.

The themes which are considered are close to the heart of the work of CIB Commissions W87 and W103, namely insurance, the role of experts and technical reports, and process and resolution.

Insurance

A number of countries give examples of cases stating that none of the parties was insured or that no insurance covered any aspect of the damage. The Type A responses made it unsurprising that all three Latin American jurisdictions, **Argentina, Chile** and **Uruguay,** were in this category; the same is true of **China.** No insurance aspects were raised in the responses from **Norway** and **Australia,** but no significance should be read into this; the Australian jurisdiction for the Type B Case Studies was New South Wales rather than Victoria, but the explanation lies no doubt in the fact that insurance was not an issue in the cases.

A second category of Type B case studies featured exclusively professional indemnity insurance of design consultants. This is true of **Canada, Quebec** and of one of the responses for **Denmark** (the other had no insurance issue) and of the two case studies from **Germany**; the type A response established that this position would obtain in **Kuwait** and sometimes in **Singapore**. The architects in **Spain** and also, in one of the studies offered, the *arquitecto técnico,* or 'project architect' would carry professional indemnity cover.

The remaining countries all reported an insurance feature of some kind worthy of note. Several countries have specialist organisations or schemes offering cover, such as the National House Builders Council (NHBC) in **England and Wales** and the Registration Organisation of Warranted Houses in **Japan** and the Home Owners' Warranty insurers in the **United States. France** offered an excellent example of the involvement of its two-stage insurance, the *dommage-ouvrage* building defects insurance and the *décennale* liability insurance of the producers; although the outcome was not a 'classic' one, in that the *dommage-ouvrage* insurer refused cover on the ground that damage caused by drought came outside its scope, and it took the court to enforce payment; an interesting contrast was offered by its two Type B Case Studies, in that the second concerns state property and would thus be exempt from the statutory requirement of *dommage-ouvrage* cover.

An interesting feature was the number of countries beyond **France** where the contractor carried insurance, a practice now taken up in some states of **Australia,** notably Victoria. The second case study from **England and Wales** involved a specialist sub-contractor and a main contractor carrying insurance, although the damage suffered fell within the latter's excess. In **Sweden,** and in both the Case Studies from **Switzerland** and the **United States,** the contractor was stated to be insured. The **United States** Type B submissions contain much helpful interpretation of insurance provisions.

Particular mention should be made of a special insurance product described in the **Hong Kong** case study of cracking of the pre-cast planks in a railway station, where the client, a government owned corporation, had comprehensive cover for design and construction which was based on an Owner Controlled Insurance Policy of the type used by the Hong Kong Government of the time on the Airport Core Programme.

The role of experts and technical reports

The main feature to emerge from an analysis of these aspects of the Type B Case Studies was the split between appointment of experts by the respective parties to a dispute and appointment in some way of an expert by the court. Nearly all the countries were familiar with the instruction of experts. In the case submitted by **China,** the parties presented their own evidence and in **Japan** there was no example of use of independent expert witnesses, although the Registration Organisation of Warranted Houses makes its own accident reports. **The United States** and **Singapore** responses did not deal expressly with experts, although they are appointed routinely in litigation in both jurisdictions. There was a divergence among those countries where the parties to disputes appointed experts, based upon the question as to who would pay for their services. In the cases from **Australia,** and from **Argentina, Canada, Quebec, Chile, Denmark, England and Wales, France, Germany, Hong Kong, Norway, Spain, Sweden, Switzerland** and **Uruguay**, at least one party and sometimes both, appointed experts to advise them on their claims. In **England and Wales** and **Norway,** to take two examples, these experts would typically become expert witnesses. Expert witnesses provide not only private guidance to the parties, but submit written evidence to the court or even appear to be examined and cross-examined by the respective sides.

It was noticeable that insurers were often instrumental in instructing experts. This could be an appointment basically to assist the insured party, as in **Canada, Quebec, Germany** and **Sweden**, but in **Denmark**, for example, an insurer instructed an expert in addition to those appointed by the parties. Unlike the Registration Organisation of Warranted Houses in **Japan,** the NHBC in **England and Wales** commissions independent reports.

Court-appointed experts are an integral system of the judicial system in **France,** although in the example given it was the plaintiff who paid for the report. In the Bechs case from **Norway,** it was again a tribunal, the Consumer Council, who commissioned a report, but again it was the plaintiffs who paid.

Court appointed experts also featured in cases from **Argentina, Denmark, Spain** and **Uruguay**. The Arbitration Court in the first case from **Germany**, could have appointed an expert but did not do so; the court appointment in the second case proved to be of little assistance. Special mention should be made of **Switzerland** where there is provision for the agreement of a *Schiedsgutachten* or binding report between the experts. In addition, an official report from a lower court was used in the litigation and the Federal Supreme Court appointed another independent expert. The *Schiedsgutachten* cannot be overruled by a judge in **Switzerland**.

Court-appointed experts have hitherto been largely unknown in the Anglo-Saxon systems in this type of case, but the NHBC case study from **England and Wales** makes reference by implication to the Woolf Report, which heralds a movement in favour of them, which, on the evidence of the Type B submissions, is the majority position.

Process and resolution

It would be wrong to attempt to adduce any criticism of legal systems or of attitudes from the case studies provided. Personalities of the parties or other circumstances may make prolonged confrontation inevitable and settlement virtually impossible. Nevertheless, the Case Studies provided do contain a wide range of possible mechanisms and outcomes for post-construction dispute resolution. Litigation was a common theme. Three variations on that theme were remarked.

First, it should not be assumed that the parties went lightly or hastily to litigation. Litigation only followed when attempts at reaching a settlement failed in case studies provided from **France**, **Germany** and **Uruguay**. In

Australia, such a failure led to arbitration. It was noted from Type A that in **Kuwait,** failure of negotiation would lead to litigation. In the **China** case study, the client agreed to a non-litigation path, but the supplier declined. The attractions of an out-of-court settlement, where this is possible, are plain to see. The huge case, one of the largest of its kind ever in **Hong Kong,** was settled, and an amicable settlement between insurers and producers was recorded in one of the Type B cases from **France** and between the disputants in an example from **Argentina.**

It is possible for one party to escape from the litigation process by settling while the others are caught up in that potentially expensive and time-consuming experience: in case studies from **Chile** and **Germany** one party managed to achieve a settlement before the others embarked upon litigation. There was evidence of formal assistance being attempted in several jurisdictions to produce settlement in place of litigation. **Canada, Quebec** offers an instructive example of an insurer initiating a mediation which helped to produce an agreement, and the air-conditioning case from **England and Wales** was resolved following a one-day mediation in front of a mediator and an air-conditioning specialist, with agreed contributions from the producers. The Building and Construction Arbitration Court instituted a conciliation procedure in **Denmark** in 1993, but its success cannot yet properly be determined. Conciliation is the order of the day according to the Case Studies from **Japan,** but it must be recalled that this is a more formalised procedure than in Western societies. In **Norway,** the government-financed Consumer Council seeks to mediate in consumer disputes.

That litigation can be a lengthy and financially painful experience can be surmised from the number of appeal hearings reported. Cases from **Chile, Denmark, Norway, Singapore, Spain, Sweden, Switzerland** and the **United States** went to appeal and only the size of the sum in dispute prevented a further appeal in **Uruguay.**

Yet it should not be assumed that litigation always has an unfortunate result. It may be that the psychological or economic effect is to concentrate the minds of the parties on a better solution, for many examples are found in the Type B Case Studies of litigation being broken off or otherwise ending in agreement. In the **United States** example, the settlement reached out of court was USD 1.6 million, Agreements were also reached during litigation in case studies from **Argentina, England and Wales, Hong Kong** and **Spain** and in an arbitration in **Australia.**

A final point worthy of note is in the examples of specialist courts and tribunals set up to deal with these often highly complex matters (and other construction disputes). The Official Referees' court in **England and**

Wales is now the Technology and Construction Court, and **Denmark** has the Building and Construction Arbitration Court, but of particular interest is the Consumer Disputes Committee in **Norway**, a quasi-judicial body financed by the Government and comprising an eleven-member expert tribunal headed by lawyers. This has been especially successful in handling smaller post-construction claims involving consumers, as its name suggests, and innovative mechanisms of this kind may well attract attention from jurisdictions where resolving such disputes can still be highly problematic.

Glossary

ARBITRATION   Usually the name given to dispute resolution by the decision of an independent third party (the Arbitrator) agreed by the disputants. Is normally an alternative to LITIGATION, but can form part of court proceedings in some countries.

CAVEAT EMPTOR   'Let the buyer beware'; a presumption that the purchaser is responsible for inspecting the subject of the purchase for any defects.

CIVIL CODE   Many of the Case Studies from civil law countries refer to this; the codification of statutory rules relating to commercial and other private law matters. Reference may also be made to a separate COMMERCIAL CODE.

COMMERCIAL CODE   The codified section of the law, chiefly in civil law systems, which deals with business matters.

CONSEQUEN-TIAL LOSS   Loss other than personal injury or damage to a person's goods or other property. Different systems may have different rules, but this would typically include the cost of making good a defective or damaged building. Economic loss resulting from breach of contract (such as the presence of a defect) is sometimes called consequential loss.

CONTRACT OF SALE
The contract by which a building (in this context) is bought and sold. This will often be different from the contract under which the building is constructed.

CONTRACTOR'S ALL RISK INSURANCE
A less than accurate name, since it is rather more concerned with risks of the contractor's liability for losses which it causes during and as a result of the construction process. It would *not* normally cover its potential liability for damage caused by post-construction defects.

DAMAGE
The harmful results of a defect. See the Green Book for further consideration.

DAMAGES
Financial compensation awarded to a party suffering loss, in this context usually as a result of breach of contract or tort (not to be confused with DAMAGE).

DAÑO MORAL
(Spanish) literally 'moral damage' a category of miscellaneous types of damage, including loss of reputation, loss of prestige, distress and disturbance.

DECENNIAL INSURANCE
Decennial insurance is intended to cover DECENNIAL LIABILITY.

DECENNIAL LIABILITY
Literally 'liability lasting ten years' but has come to refer to a range of related liability provisions, especially under the French and other Napoleonic systems (*décennale*).

DEFECT
Departure from, or disconformity with the requirements as to quality imposed by contract or by regulation.
See Green Book for further consideration.

DOMMAGE-OUVRAGE	An important element of the French insurance system whereby payment can be made to the owner to finance repairs before the producers' liability is established. The *dommage-ouvrage* insurer has recourse against the appropriate liability insurer.
DUTY OF CARE	A duty limited to the use of care (often 'reasonable care and skill' or similar formulation) rather than the higher duty of result (See DUTY OF RESULT)
DUTY OF RESULT	A duty higher than that of the duty of care, being a duty to achieve a particular result, standard or quality. Similar to the concept of FITNESS FOR PURPOSE.
ECONOMIC LOSS	Loss other than personal injury or damage to a person's goods or other property. Different systems may have different rules, but this would typically include the cost of making good a defective or damaged building. Economic loss resulting from breach of contract (such as the presence of a defect) is sometimes called consequential loss.
EXTRA-CONTRACTUAL	Literally 'outside of contract' and could mean outside an individual contract. Alternatively, could refer to legal rules and rights of action beyond the Law of Contract: thus tort could be said to give extra-contractual rights.
FEEDBACK	In the context of this book, the collection of data on failures and claims, for reporting back to insurers, regulatory bodies, building research establishments and others.

FITNESS FOR PURPOSE	The standard of goods for a building which means that they can operate or be used as intended. See DUTY OF RESULT.
GREEN BOOK	'Post Construction Liability and Insurance', ed. J. Knocke (1993), E&FN Spon, London (ISBN 0–419–15350–0).
IN SOLIDUM	Liability considered jointly. See JOINT AND SEVERAL LIABILITY and PROPORTIONATE LIABILITY.
JOINT AND SEVERAL LIABILITY	The principle that, where more than one party is liable in respect of damage caused, each can be wholly liable for the total. See PROPORTIONATE LIABILITY and *IN SOLIDUM*.
LATENT DEFECT	A defect in the building which is not detectable by reasonable inspection until it manifests itself in damage.
LATENT DEFECTS INSURANCE	Insurance intended to cover the risk of damage occurring after construction as a result of latent defects. A similar concept to *DOMMAGE-OUVRAGE* but with significant differences. BUILD (Building Users Insurance Against Latent Defects) is an example of such insurance.
LITIGATION	The hearing of legal cases in the courts.
MEDIATION	A method of dispute resolution which is an alternative to LITIGATION and ARBITRATION, by which an independent mediator is brought in to try to assist the parties in reaching a settlement.

POST-CON-STRUCTION LIABILITY	This refers to the period after construction, often starting with a fixed point, such as *réception* in France, or 'hand-over', but the point will vary from system to system.
PRESCRIPTION	Although in law a different concept from limitation, the effect is similar, namely that at the end of the prescription period, an action cannot be brought.
PRIVITY OF CONTRACT	The rule that only parties to a contract can sue or be sued under that contract.
PRODUCER	Equivalent to the French term *constructeur* and the Spanish *constructor*, encompasses those engaged in the supply or construction of a building e.g. contractor, sub-contractor, architect, engineer. See the Green Book Chapter 3 and Appendix B.
PROFES-SIONAL LIABILITY INSURANCE or PROFES-SIONAL INDEMNITY INSURANCE	Insurance of the risk of loss arising from the carrying out of professional duties.
PROPORTION-ATE LIABILITY	The principle that producers should only incur liability for that proportion of the work for which they were responsible. Is a major departure from JOINT AND SEVERAL or *IN SOLIDUM* LIABILITY.

QUANTI *MINORIS* ACTION	Action in law claiming a reduction in the price paid, for example, because of the defects in a building.
REDHIBITORY ACTION	An action claiming entitlement to rescind a contract because of a breach.
RUINA	(Spanish) Would originally have meant something close to 'ruin' in English, i.e. serious damage to a building amounting actually or virtually to its destruction, but Spanish jurisprudence (for example) has expanded the meaning greatly to cover 'unfunctionality of the building' in many senses.
WARRANTY	A legally enforceable promise or assurance which may form a contract or part of one and which may or may not be transmissible between parties e.g. vendor and purchaser.

The following terms are considered in the Green Book, in Part One and/or in Appendix B:

General Terms: Client, Damages, Hand-Over, Post-Construction, Producer, Successive owners, Tenant

Aetiological Terms: Breach of Duty, Damage, Defect, Error or Omission

Insurance Terms: Claim(s) Event, Excess, Franchise, Insured Party, Policy Holder

Currencies

ARS	Argentina (peso)	CLP	Chile (peso)
AUD	Australia (dollar)	CNY	China (yuan)
CAD	Canada (dollar)	DEM	Germany (deutsche Mark)
CHF	Switzerland (franc)	DKK	Denmark (krone)
		ESP	Spain (peseta)

FRF France (franc)
GBP United Kingdom (pound)
HKD Hong Kong (dollar)
JPY Japan (yen)
KWD Kuwait (dinar)

NOK Norway (krone)
SEK Sweden (krona)
SGD Singapore (dollar)
USD USA (dollar)
UYU Uruguay (new peso)

Subject Index

For Product Safety Concerns and Information please contact our EU
representative GPSR@taylorandfrancis.com
Taylor & Francis Verlag GmbH, Kaufingerstraße 24, 80331 München, Germany

www.ingramcontent.com/pod-product-compliance
Ingram Content Group UK Ltd.
Pitfield, Milton Keynes, MK11 3LW, UK
UKHW021622240425
457818UK00018B/687

* 9 7 8 0 3 6 7 3 9 9 6 4 1 *